Ecosystem Services

ISSUES IN ENVIRONMENTAL SCIENCE AND TECHNOLOGY

How to obtain future titles on publication

A subscription is available for this series. This will bring delivery of each new volume immediately on publication and also provide you with online access to each title via the Internet. For further information visit http://www.rsc.org/Publishing/Books/issues or write to the address below.

For further information please contact:
Sales and Customer Care, Royal Society of Chemistry, Thomas Graham House, Science Park, Milton Road, Cambridge, CB4 0WF, UK
Telephone: +44 (0)1223 432360, Fax: +44 (0)1223 426017, Email: sales@rsc.org

ISSUES IN ENVIRONMENTAL SCIENCE AND TECHNOLOGY

EDITORS: R.E. HESTER AND R.M. HARRISON

30
Ecosystem Services

RSC Publishing

ISBN: 978-1-84973-018-1
ISSN: 1350-7583

A catalogue record for this book is available from the British Library

Published by The Royal Society of Chemistry,
Thomas Graham House, Science Park, Milton Road,
Cambridge CB4 0WF, UK

Registered Charity Number 207890

For further information see our web site at www.rsc.org

Preface

The rapid increase in world population coupled with an ever-increasing demand for natural resources, including fresh water, has highlighted the immense pressures under which the environment is placed. This led initially to an emphasis on technological solutions, such as end-of-pipe clean-up of effluents and increases in agricultural production through plant breeding and intensive use of chemical fertilisers and pesticides. There has been a progressive recognition that such approaches are not sustainable in the long term and this has led to more holistic thinking about environmental management. The outcomes of this changed thinking include, for example, a life cycle approach to the assessment of manufactured products accounting for mass and energy flows at all points in the process.

Even these newer modes of thinking have failed to account adequately for pressures on the environment. Society has tended to regard the environment as a 'free good' which could simply be exploited without concern for the consequences. There was recognition that this would imply loss of some species as a result, but the real consequences of this loss were not fully appreciated. It is only with the more recent development of thinking in terms of ecosystems services that the full impact of human activities is beginning to be recognised. When human settlements were relatively small, it was possible to discharge effluent into local water courses which would render the waste innocuous through self-purification processes, now recognised as an ecosystem service. This, however, is only one example of very many ecosystem services which extend through all environmental media and underpin the very basis of human existence. Our most basic needs for food, water and air are supplied through ecosystem services which society interferes with at its peril.

The first chapter in this volume, by Alastair Fitter and other members of the European Academies Science Advisory Council (EASAC) assessment team, outlines some of the key ecosystem services, assesses their importance in a European context and examines the contribution which biodiversity makes to each of them. A key insight from this authoritative study is that all ecosystems deliver a broad range of services, and managing an ecosystem primarily to

Issues in Environmental Science and Technology, 30
Ecosystem Services
Edited by R.E. Hester and R.M. Harrison
© Royal Society of Chemistry 2010
Published by the Royal Society of Chemistry, www.rsc.org

deliver one service will reduce its ability to provide others. This is amplified through examining the use of land to produce biofuels, a superficially attractive process, which in practice, is riddled with difficulties and disadvantages. The following two chapters take specific ecosystems as examples. Piran White and co-authors review coastal wetland ecosystem services and the many benefits which they provide to humans including nutrient cycling, climate and water quality regulation, timber fuel and fibre. A new integrative conceptual framework to underpin the ecosystem approach is divided into three sub-systems relating to ecosystem functions, ecosystem services and social development and well-being. The subsequent chapter by Ken Norris, Simon Potts and Simon Mortimer deals with ecosystem services and food production. This is an absolutely key area with the world's population projected to rise to eight billion people and world food demand increasing by 50% by the year 2030. An overview is provided of the key ecosystem services involved in different food production systems including crop and livestock production, agriculture and the harvesting of wild nature. The important ecosystem impacts of food production systems include habitat loss and degradation, changes to water and nutrient cycles, and biodiversity loss. It is explained how these often impact upon the very ecosystem services on which food production systems depend, as well as other ecosystem services unrelated to food. More sustainable ways forward are proposed.

In the subsequent chapter, John Thornes reviews the services provided by the atmosphere. Many of these do not have a strong ecological dimension but nonetheless management of the quality and sustainability of the atmosphere is of very high importance as it represents one of the planet's key life support systems. The chapter reviews the services provided to society by the atmosphere and presents provisional economic costings of some of these which serve to emphasise the immense importance of the atmosphere to society, a fact which is often neglected. The theme of economic valuation is taken up in a more general context by Eric Gómez Baggethun and Rudolf De Groot in the following chapter on 'Natural Capital and Ecosystem Services: The Ecological Foundation of Human Society'. This chapter highlights the problem that standard economic theory neglects, the fact that economic health in the long term depends upon the maintenance of the integrity and resilience of natural ecosystems which underpin so much of human activity. It describes how approaches such as ecological and environmental economics attempt to deal with the short-comings of standard economics through the development of concepts and accounting methods that better reflect the role of nature in the economy and the ecological cost derived from economic growth.

The final two chapters deal with practical ways of addressing sustainability issues which affect ecosystems. Luke de Vial, Fiona Bowles and P. Julian Dennis provide a case study of water management. Specifically they examine how agricultural practices impact upon water quality and require ever more complex and expensive measures to remediate water for potable supplies. Their case study describes tackling the problem at source by influencing the farmers whose practices are at the root cause of the problem. It is very encouraging to see that such methods are meeting with success. In the final chapter, Adisa Azapagic describes the use of life cycle assessment as a tool for sustainable management of ecosystem services. The theoretical basis of life cycle analysis is described, but the key part of the chapter is the presentation of four case studies, which illustrate the insights which can be gained from life cycle analysis into the comparative impact of different process options and fuels upon the environment.

This is a relatively new subject area and it has proved difficult to provide the kind of comprehensive integrated coverage which we aim for in *Issues in Environmental Science and Technology*. Rather, the volume provides a collection of specialist accounts of different aspects of ecosystem services which complement one another, and although not providing a comprehensive overview, give an excellent insight into the subject, its importance and the way in which it is developing. Each chapter is designed to be read as a complete entity in its own right, but a much fuller view of the field in terms of the scientific foundations, management approaches and policy aspects can be obtained by reading the entire volume. We are fortunate to have attracted highly authoritative authors for the individual chapters and believe that this volume will be of immediate and lasting value to the many people involved in the science and policy aspects of environmental management and sustainability in central and local government, consultancies and industry, as well as to students of environmental science, engineering and management courses.

Ronald E. Hester
Roy M. Harrison

Contents

Issues in Environmental Science and Technology, 30
Ecosystem Services
Edited by R.E. Hester and R.M. Harrison
© Royal Society of Chemistry 2010
Published by the Royal Society of Chemistry, www.rsc.org

Editors

Ronald E. Hester, BSc, DSc (London), PhD (Cornell), FRSC, CChem

Ronald E. Hester is now Emeritus Professor of Chemistry in the University of York. He was for short periods a research fellow in Cambridge and an assistant professor at Cornell before being appointed to a lectureship in chemistry in York in 1965. He was a full professor in York from 1983 to 2001. His more than 300 publications are mainly in the area of vibrational spectroscopy, latterly focusing on time-resolved studies of photoreaction intermediates and on biomolecular systems in solution. He is active in environmental chemistry and is a founder member and former chairman of the Environment Group of the Royal Society of Chemistry and editor of 'Industry and the Environment in Perspective' (RSC, 1983) and 'Understanding Our Environment' (RSC, 1986). As a member of the Council of the UK Science and Engineering Research Council and several of its sub-committees, panels and boards, he has been heavily involved in national science policy and administration. He was, from 1991 to 1993, a member of the UK Department of the Environment Advisory Committee on Hazardous Substances and from 1995 to 2000 was a member of the Publications and Information Board of the Royal Society of Chemistry.

Roy M. Harrison, BSc, PhD, DSc (Birmingham), FRSC, CChem, FRMetS, Hon MFPH, Hon FFOM

Roy M. Harrison is Queen Elizabeth II Birmingham Centenary Professor of Environmental Health in the University of Birmingham. He was previously Lecturer in Environmental Sciences at the University of Lancaster and Reader and Director of the Institute of Aerosol Science at the University of Essex. His more than 300 publications are mainly in the field of environmental chemistry, although his current work includes studies of human health impacts of atmospheric pollutants as well as research into the chemistry of pollution phenomena. He is a past Chairman of the Environment Group of the Royal Society of Chemistry for whom he has edited 'Pollution: Causes, Effects and Control' (RSC, 1983; Fourth Edition, 2001),

'An Introduction to Pollution Science' (RSC, 2006) and 'Principles of Environmental Chemistry' (RSC, 2007). He has a close interest in scientific and policy aspects of air pollution, having been Chairman of the Department of Environment Quality of Urban Air Review Group and the DETR Atmospheric Particles Expert Group. He is currently a member of the DEFRA Air Quality Expert Group, the DEFRA Expert Panel on Air Quality Standards, and the Department of Health Committee on the Medical Effects of Air Pollutants.

Contributors

Adisa Azapagic, *School of Chemical Engineering and Analytical Science, The University of Manchester, PO Box 88, Sackville Street, Manchester M60 1QD, UK*

Fiona Bowles, *Wessex Water, Claverton Down Road, Bath, BA2 7WW*

Rudolf De Groot, *Environmental Systems Analysis Group, Wageningen University, Wageningen, The Netherlands*

Luke de Vial, *Wessex Water, Claverton Down Road, Bath, BA2 7WW*

P Julian Dennis, *Wessex Water, Claverton Down Road, Bath, BA2 7WW*

Thomas Elmqvist, *Stockholm University, Sweden*

Alastair Fitter, *University of York, UK*

Jasmin A. Godbold, *Environment Department, University of York, Heslington, York YO10 5DD*

Erik Gomez-Baggethun, *Social-Ecological Systems Laboratory, Department of Ecology, Universidad Autónoma de Madrid, Madrid, Spain*

Roy Haines-Young, *University of Nottingham*

Alison R. Holt, *Environment Department, University of York, Heslington, York YO10 5DD*

Simon R. Mortimer, *Centre for Agri-Environmental Research, School of Agriculture, Policy and Development, New Agriculture Building, University of Reading, Whiteknights, PO Box 217, Reading, Berkshire RG6 6AH*

John Murlis, *EASAC, UK*

Ken Norris, *Centre for Agri-Environmental Research, School of Agriculture, Policy and Development, New Agriculture Building, University of Reading, Whiteknights, PO Box 217, Reading, Berkshire RG6 6AH*

Marion Potschin, *Nottingham University*

Simon G. Potts, *Centre for Agri-Environmental Research, School of Agriculture, Policy and Development, New Agriculture Building, University of Reading, Whiteknights, PO Box 217, Reading, Berkshire RG6 6AH*

Andrea Rinaldo, *Ecole Polytechnique FederaleLausanne, Switzerland*

Heikki Setälä, *University of Helsinki, Finland*

Martin Solan, *Environment Department, University of York, Heslington, York YO10 5DD*

Susanna Stoll-Kleemann, *University of Greifswald, Germany*

John Thornes, *School of Geography, Earth and Environmental Sciences, University of Birmingham, Edgbaston, Birmingham B15 2TT, UK*

Piran C. L. White, *Environment Department, University of York, Heslington, York YO10 5DD*
Jessica Wiegand, *Environment Department, University of York, Heslington, York YO10 5DD*
Martin Zobel, *Tartu University, Estonia*

An Assessment of Ecosystem Services and Biodiversity in Europe

ALASTAIR FITTER, THOMAS ELMQVIST, ROY HAINES-YOUNG,
MARION POTSCHIN, ANDREA RINALDO, HEIKKI SETÄLÄ,
SUSANNA STOLL-KLEEMANN, MARTIN ZOBEL AND
JOHN MURLIS[*]

ABSTRACT

Ecosystem services are the benefits humankind derives from the workings
of the natural world. These include most obviously the supply of food,
fuels and materials, but also more basic processes such as the formation
of soils and the control and purification of water, and intangible ones
such as amenity, recreation and aesthetics. Taken together, they are
crucial to survival and the social and economic development of human
societies. Though many are hidden, their workings are now a matter of
clear scientific record. However, the integrity of the systems that deliver
these benefits cannot be taken for granted, and the process of monitoring
them and of ensuring that human activity does not place them at risk is an
essential part of environmental governance, not solely at a global scale
but also regionally and nationally.

In this chapter, we assess the importance of ecosystem services in a
European context, highlighting those that have particular importance for
Europe, and we set out what is known about the contribution biodiversity
makes to each of them. We then consider pressures on European ecosystem services and the measures that might be taken to manage them.

One of the key insights from this work is that all ecosystems deliver a
broad range of services, and that managing an ecosystem primarily to

[*]Corresponding Author

Issues in Environmental Science and Technology, 30
Ecosystem Services
Edited by R.E. Hester and R.M. Harrison
© Royal Society of Chemistry 2010
Published by the Royal Society of Chemistry, www.rsc.org

deliver one service will reduce its ability to provide others. A prominent current example of this is the use of land to produce biofuels. There is an urgent need to develop tools for the effective valuation of ecosystem services, to achieve sustainable management of the landscape to deliver multiple services.

1   Introduction

1.1   *Biodiversity and Ecosystem Services: Why this Topic Matters Now*

The past 50 years have seen an unprecedented human impact on ecosystems and on their biodiversity.[1] Current rates of species extinction substantially exceed background extinction rates: International Union for Conservation of Nature (IUCN) estimates that 12% of bird species, 23% of mammals, 32% of amphibians and 25% of conifers are threatened with extinction.[2] Human use of natural resources has grown substantially in this period: roughly half of useable terrestrial land is now devoted to grazing livestock or growing crops. That expansion has been at the expense of natural habitat, so that between a quarter and a half of all primary production is now diverted to human consumption.[3] Other major threats to biodiversity include the introduction of non-indigenous species, pollution, climate change and over-harvesting. In marine ecosystems, over-exploitation of stocks has been the most severe cause of ecosystem degradation and local extinction.[4]

These changes have considerable implications for human society. Living organisms, interacting with their environment in the complex relationships that characterise ecosystems, deliver important, and in some cases crucial and unsubstitutable, benefits to humankind. Most obviously, organisms provide goods in the form of food, fuel and materials for building, but they also deliver other, less apparent services. For example, insects, especially bees, play an important role in the pollination of plants, including staple food crops, and micro-organisms recycle or render harmless the waste produced by human society. Both the bees and the microbes operate within and rely on ecosystems for their survival.

These natural services are of enormous value to human society. Many of the services are irreplaceable: for example, we have no way of providing food for the human population except through the use of natural systems involving soil, soil organisms and crop plants, nor of providing drinking water, except through the operation of the water cycle, which depends critically on the activities of organisms. The maintenance of ecosystems, therefore, must be an essential part of the survival strategy for human societies.

Despite these benefits, investment in conservation does not match the scale of the benefits received from ecosystem services. It was noted by David Pearce that 'actual expenditures on international ecosystem conservation appear to be remarkably small and bear no relationship to the willingness to pay figures obtained in the various stated preference studies'.[5] Pearce concluded 'despite all the rhetoric, the world does not care too much about biodiversity conservation'. This disconnection may arise in part because the links between biodiversity and ecosystem function (and consequently to ecosystem services) remain new areas of research: this chapter assesses the evidence for these links, focussing on ecosystem services that are of major concern for Europe.

The power of economic analysis in policy-making is such that argument about policy is typically constructed in a major part through the language of costs and benefits. There is an urgent need to address the chronic under-investment in conservation of biodiversity and to ensure that future decisions do not lead to an unacceptable loss. This means that it is essential that the value of biodiversity in promoting the delivery of essential and valuable services is expressed strongly (in both economic and other terms) in those areas of decision-making where economic analysis is itself strongest.

1.2 The Current Assessment

The principal focus of assessment of ecosystem services to date has been at a global level. The Millennium Ecosystem Assessment (MA) continues to be a major influence on the development of a global regime for the protection of biodiversity through the Convention on Biological Diversity (CBD). At a national scale, UK National Ecosystem Assessment (NEA), which commenced in mid-2009 and will report in 2011, is expected to have a significant impact on the UK's environmental management strategy. There is also an urgent need to advance the development of regional measures for protecting biodiversity and ensuring the continual flow of ecosystem services. The assessment on which this chapter is based was commissioned by the Council of the European Academies Science Advisory Council (EASAC), an independent association of the science academies of the European Member States, as a contribution to the scientific debate on the future of European biodiversity and measures to protect it.[6]

The assessment consists of four stages:

1. Prioritisation of ecosystem services within a European context using the MA framework;
2. Assessment of the relative significance of biodiversity for each of these services;
3. An evaluation of the role of biodiversity, based on current knowledge; and
4. Identification of specifically European concerns about the future of each ecosystem service.

The initial assessment was made by an expert Working Group. Following extensive review by a wide range of experts, comments and contributions from reviewers were assimilated and the output was subject to a review within the EASAC Member Academies. We believe, therefore, that this assessment is an accurate reflection of the range of views within Europe's scientific communities on ecosystem services and biodiversity in Europe.

2 Biodiversity and Ecosystem Services

2.1 Ecosystem Services

An ecosystem is the interacting system of living and non-living elements in a defined area.[7] Ecosystems can exist at any spatial scale, although in most uses they are large-scale entities, such as a lake or a forest. The importance of the ecosystem is that it is the level in the ecological hierarchy (see Figure 1) at which key processes such as carbon, water and nutrient cycling and productivity are determined and can be measured: these are the processes that determine how the world functions and that underlie all the services identified by the MA.

The MA classification of ecosystem services contains four categories – supporting, regulating, provisioning and cultural – which explicitly address the benefits to human societies. The delivery of these services, however, represents the normal operation of the ecosystem, and reflects the natural processes that occur within every ecosystem. The services, therefore, which are a human construct, depend on these underlying processes, such as:

- Fixation of nitrogen gas from the air by bacteria into forms that are useable by plants, which underlies the nitrogen cycle;
- Decomposition of organic matter by microbes, which is the basis of all nutrient cycles, including importantly the carbon cycle; and
- Interactions between organisms, such as competition, predation and parasitism, which control the size of their populations, and underlie services such as pest control.

Because the processes depend on organisms and the organisms are linked by their interactions, the services themselves are also linked. For example, productivity can only be maintained if the cycling of nutrients continues, and all provisioning services depend intimately on the supporting services of production and water and nutrient cycling. Consequently all ecosystems deliver multiple services, although the number of species and the relative scale of the various services will vary greatly among ecosystems.

2.2 Relationships between Biodiversity and Ecosystem Services

Ecosystems vary greatly in biodiversity. Generally, productive natural ecosystems have the highest biodiversity but many highly productive ecosystems, and especially those under human management, have low biodiversity, showing that many other factors are at work. Among those factors are: rates of evolution, which are the underlying driver of biodiversity; rates of dispersal, both natural and assisted by humans, which are especially important when

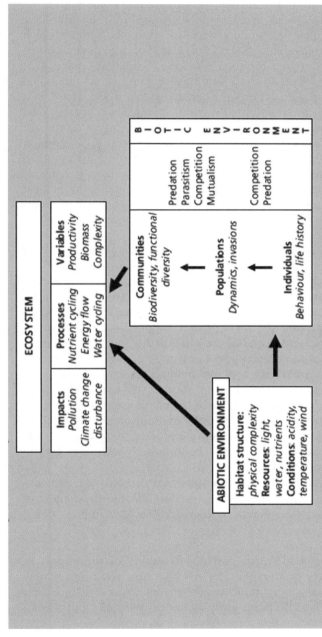

Figure 1   The Ecological Hierarchy. (Reproduced with permission of EASAC, taken from the original report on ecosystem services prepared by Alastair Fitter and the EASAC working group.)

ecosystems are isolated from others by natural barriers; and the interactions between species, such as predation, competition and parasitism, which control the sizes of their populations and often their persistence in a community.

In ecosystems with many species, species can be grouped into sets that have similar ecological roles, called functional groups, for example, legumes which form a symbiosis with nitrogen-fixing bacteria in their roots and gain access to the pool of atmospheric nitrogen for their nutrition. Similarly, spiders that catch prey in webs and those that do so by hunting represent distinct functional groups of predators and play distinct roles in an ecosystem. Even where, in a biodiverse ecosystem, there are many species within a functional group, some will be rare and others common. Some species play especially important roles in the ecosystem, although these keystone species may not necessarily be common species. Losing an entire functional group from an ecosystem or the keystone species from within that group is likely to have more severe consequences for its functioning than losing one species from a large group, and such a loss is most likely in a species-poor system.[8] Experimental evidence shows that both number of species and number of functional groups can play an important role in controlling ecosystem processes.[9]

Ecosystems can change drastically when sets of key species are lost,[10,11] or when new species invade.[12] One of the great unsolved problems in ecology is to determine how important biological richness is for the operation of processes such as production and nutrient cycling. When there are more species in an ecosystem, and especially more types of species with distinct functional attributes, ecosystem processes, and the services they support, such as biomass production, pollination and seed dispersal are promoted,[13] but the evidence is less clear as to what happens to an ecosystem as it progressively loses species. Because processes in ecosystems with very low biodiversity are in many cases slower or less active, it follows that loss of species will eventually cause degradation of processes. Although the shape of the relationship is not entirely clear (do services decline progressively or suddenly as biodiversity is lost?) there is evidence that it is highly non-linear. A slight decreasing trend in ecosystem functions as species diversity declines may be followed beyond a certain threshold with a collapse of function.[14]

There are numerous well-documented examples that demonstrate that biodiversity plays a large role in the case of many services. Within the context of the Millennium Ecosystem Assessment framework, such examples would include:

- *Supporting services*: in a meta-analysis of 446 studies of the impact of biodiversity on primary production, 319 of which involved primary producer manipulations or measurements, there was 'clear evidence that biodiversity has positive effects on most ecosystem services', and specifically that there was a clear effect of biodiversity on productivity.[15]
- *Regulating services*: in an experimental study of pollination in pumpkins it was the diversity of pollinator species, and not their abundance, that determined seed set.[16]

- *Provisioning services*: where grassland is used for biofuel or other energy crop production, the lower financial return makes intensive production systems involving heavy use of pesticides and fertilisers uneconomic. Under these less intensive production systems, mixed swards of grasses are more productive than pure swards.[17]
- *Cultural services*: evidence from the 2001 foot and mouth disease epidemic in the UK demonstrated that the economic value of biodiversity-related tourism greatly exceeds that of agriculture in the uplands of the UK.

2.3   Land Use and Multiple Services

Land use has a major impact on both ecosystem services and biodiversity, especially when altered by human activity to deliver some particular service, such as food production in agro-ecosystems. However, all ecosystems deliver multiple services, and management to maximise one particular service risks reducing others. For example, forests regulate water flow and quality and store nutrients in soil, among many other functions; clear-felling a forest to obtain the ecosystem service of timber products results in the temporary failure of the system to retain nutrients, as shown by the classic Hubbard Brook experiments in New England, USA.[18] Similarly, arable land managed to maximise yield of food crops stores less carbon in the soil, with negative effects on the service of climate regulation.[19]

Human impact on ecosystems is most extreme in intensive agriculture and in urban landscapes. Urban ecosystems typically contribute minimal levels of provisioning services. Urban landscapes are characteristically heterogeneous, including in relation to biodiversity.[20] Street trees and urban vegetation may generate services of high value for human well-being related to environmental quality such as air cleaning, noise reduction and recreation,[21] or to human health[22] (asthma rates among children aged four and five in New York City were directly proportional to the density of trees). Because of the density of human population, many urban ecosystem services are generated on a very small scale, by patches of vegetation and even individual trees.

Land (and where appropriate water) management always, even if only implicitly, aims to achieve benefits of one or more ecosystem services, but because these services are not independent of one another, there are trade-offs between the services.

- *Temporal trade-offs*: there may be benefits now with costs incurred later (or more rarely *vice versa*). Land used for food production may store progressively declining stocks of organic matter, with long-term consequences both for nutrient cycling, and hence future fertility, and carbon sequestration.
- *Spatial trade-offs*: the benefit may be experienced at the site of management, but the cost incurred elsewhere. When moorland is burned to maximise growth of young heather shoots and the number of grouse, and hence income from grouse shooting, the loss of dissolved organic matter to

water is increased. This appears as colour in drinking water and has to be removed at great expense by water companies.[23]

- *Beneficiary trade-offs*: the manager may gain benefit, but others lose, leading to actual or potential conflict. Most management systems that maximise production by high inputs of fertilisers lead to reduced biodiversity, so that those who appreciate land for its conservation value lose. Equally, land managed for biodiversity conservation, such as nature reserves, has little production value.
- *Service trade-offs*: these occur almost invariably when management is principally for one service, and are in practice similar to beneficiary trade-offs.

These trade-offs are real and well documented. To control their impact, it will be essential to take into account the spatial and temporal scale at which ecosystem services are delivered. For example:

- Pollination, which operates at a local scale and can be managed by ensuring that there are areas of land managed that maintain populations of pollinators in a mosaic of land-use types;
- Hydrological services function at a landscape scale, such as a watershed, and require co-operation among land managers at that scale; and
- Carbon sequestration in organic matter in soil operates at a regional and global scale and necessitates policy decisions by governments and international bodies to ensure that appropriate incentives are in place to ensure necessary behaviour by local land managers.

Hence the importance of assessments at a range of geographical scales, including, as in the work reported here, regional (European in this case) level.

3   European Biodiversity and Ecosystem Services

The full assessment of ecosystem services made in the course of this study by the EASAC Working Group and the panel of experts is given in the Working Group Report.[24] The following is a digest focussing on land-based services; similar issues are raised by consideration of marine services.

A Supporting Services
Supporting services are the basic services that make the delivery of all other services possible.

A1 Primary Production
Primary production is fundamental to all other ecosystem services and is generally high in Europe, where soils are young and fertile and the climate is generally benign. Low productivity is associated with very cold regions (Arctic and alpine), very dry regions (some parts of the Mediterranean region) and seriously polluted or degraded environments. In policy terms, primary

production is considered highly important for Europe, and it appears to be strongly dependent on biodiversity.

A large body of evidence relates diversity to primary production, including theoretical, controlled-environment and small and large-scale field studies. However, the relationship is complex. Although the highest productivity is typically achieved in intensively managed systems of very low diversity requiring large inputs of resource, sustained high production without high levels of input is associated with high levels of biodiversity.

In Europe, there is a possibility of serious decline in primary productivity due to increasingly dry conditions in southern Europe, but of increases in the north due to extended growing season. Environmental pressure, including change of land use, climate change and pollution all reduce quantity and quality of biodiversity with consequent loss of primary productivity.[25]

A2 Nutrient Cycling
Nutrient cycling is also considered a highly important ecosystem service for Europe. It is a key process in both terrestrial and aquatic systems and is essential for maintenance of soil fertility. Nutrients are cycled by organisms, which take them up as they grow and release them back into the environment as they decompose. Biodiversity is critical to these cycles.

The capacity of ecosystems to sequester nutrients depends, besides natural factors, on management interventions. In intensively farmed landscapes, nitrate and phosphate may be lost to watercourses, causing both damage to water quality and economic losses on farms. Atmospheric deposition of nitrogen, sulfur and sometimes metals to soils also disrupts nutrient cycling – through effects including acidification, denitrification and inhibition of fixation. In many aquatic systems in Europe, sewage, industrial and agricultural effluent disrupt nutrient cycling.

The widespread use of sewage sludge as an agricultural fertiliser, though an effective way of recycling nutrients removed from soils by agriculture, has resulted in contamination of soils by heavy metals, including zinc, copper and cadmium, which inhibit nitrogen-fixing bacteria. Changes in biodiversity of natural ecosystems brought about by land-use change, climate change or pollution alter the ability of ecosystems to retain nutrient stores, resulting in release of nutrients to other ecosystems with potentially damaging consequences.

A3 Water Cycling
The water cycle is an important process in the overall management of water, storing water, controlling flows and distributing it to all parts of the ecosystem. Humans have made changes in water cycles through urbanization, drainage, dams, structural changes to rivers and other surface waters.[26] Floods and droughts become more intense due to changes in landscapes and feedbacks from precipitation recycling, which include forest cutting, intensive agriculture, urbanization, large-scale reclamations and uncontrolled withdrawals from subsurface stores.[27] Impermeable areas increasingly preclude sustainable aquifer recharge. Impacts are likely to be amplified through climate change.[28]

Both vegetation and soil organisms have profound impacts on water movements and the extent of biodiversity is likely to be important. Changes in species composition can affect the balance between water used by plants ('green water') and water flowing through rivers and other channels ('blue water'), and native flora may be more efficient at retaining water than exotic species. However, land use and landscape structure are likely to be more significant than biodiversity *per se*.

In Europe as a whole, there is concern that soil moisture and green water availability are decreasing as a result of human activity[29] and in Southern Europe these problems apply to both blue and green water. Urban areas with sealed surfaces provide new challenges and increased runoff, flood events and nonpoint pollutant loads[30] are predicted to increase in several European areas due to climate change.[31]

A4 Soil Formation

Soil formation is fundamental to soil fertility, especially where processes leading to soil destruction or degradation (erosion or pollution) are active. It is a continuous process in all terrestrial ecosystems, but particularly important and active in early stages after land surface is exposed (*e.g.* following glaciation). It is highly dependent on the nature of parent materials, biological processes, topography and climate.

Soil biodiversity is a major factor in soil formation. Loss of soil biota, including bacteria, fungi and invertebrates, reduces soil formation rate, with damaging consequences. Key plant types include legumes and deep-rooted species. There is little empirical evidence, however, on the general role of biodiversity in soil formation, but composition of biological communities has been shown to be important, so a range of functional types appears to be needed.

There are particular concerns in Europe about soils that are subject to intense erosion by wind or water. Although soils in Northern European ecosystems in the early stages (10 000 to 20 000 years) of post-glacial recovery are often resilient to intensive agricultural use,[32] much of the Mediterranean region has old soils with lower resilience that have suffered severe damage and are badly eroded.[33] In alpine areas, high rates of erosion may be countered by equally high rates of soil development.

B Regulating Services

Regulating services are the benefits obtained from the regulation of ecosystem processes.

B1 Climate Regulation

Climate regulation refers to the role of ecosystems in managing the levels of climate forcing or greenhouse gases (GHGs) in the atmosphere. Current climate change is largely driven by increase in the concentration of trace gases in the atmosphere, principally as a result of changes in land use and rapidly rising combustion of fossil fuels. The major GHG, carbon dioxide (CO_2), is absorbed directly by water and indirectly (*via* photosynthesis) by vegetation, leading to storage in biomass and in soils as organic matter. Fluxes of other GHGs (*e.g.* methane, CH_4; nitrous oxide, N_2O) are also regulated by soil microbes. Marine systems play a key role in climate regulation through physical absorption of CO_2 and through photosynthetic carbon-fixation.

Europe contains extensive areas of peat that contain large quantities of carbon. Boreal forests are also significant stores of carbon. In all, Europe's terrestrial ecosystems are estimated to represent a net carbon sink of between 135 and 205 gigatonnes per year, which is about 7 to 12% of the 1995 anthropogenic carbon emissions.[34] The interplay between biodiversity and climate regulation is poorly understood. When major change occurs in ecosystems, the time lags in the feedbacks on ecosystem processes that result are important and unresolved. The global carbon cycle is strongly buffered because much anthropogenic CO_2 is absorbed by the oceans and terrestrial ecosystems.[35] However, the rate of emission increasingly exceeds this absorption capacity, which itself is being reduced still further by anthropogenic damage to ecosystem function.

Losses of carbon (C) from soils, from peat in particular, could easily outweigh any savings made due to reductions in fossil fuel use: it has been estimated that UK soils may have lost 0.6% of C each year over last 25 years.[36] Intensive biofuel production may also lead to reduced C retention in soils, since the goal will be to remove as much biomass as possible. There is also some evidence that aerosols produced by boreal forests may affect albedo, thereby cooling the climate.[37]

There is a fundamental requirement to ensure that policies take into account multiple impacts; for example, the consequences of changes in land use to increase biomass production for sustainable C storage in soils and emissions of greenhouse gases (N_2O, CH_4).

B2 Disease and Pest Regulation

The abundance of pests and diseases is regulated in ecosystems through the actions of predators and parasites, as well as by the defence mechanisms of their prey. The services of regulation are expected to be more in demand in the future, as climate change brings new pests and increases the susceptibility of species to parasites and predators.

The role of biodiversity in disease regulation may be important. There is evidence that the spread of pathogens is less rapid in more biodiverse ecosystems. There is also a consensus that a diverse soil community will help prevent losses of crops due to soil-borne pests and diseases.[38] Higher trophic levels in soil communities can play a role in suppressing plant parasites and affecting nutrient dynamics by modifying abundance of intermediate consumers.[39] In many managed systems, the control of plant pests can be provided by generalist and specialist predators and parasitoids.[40,41]

There is a need for the development of European applications of biological control, exploiting the properties of pest regulation in biodiverse ecosystems.

B3 + C2 Water Regulation and Purification (in this assessment these ecosystem services were combined)

The water regulation and purification service refers to the maintenance of water quality, including the management of impurities and organic waste, and the direct supply of clean water for human and animal consumption. Soil state and vegetation both act as key regulators of water flow and storage. Although vegetation is a major determinant of water flows and quality, and micro-organisms play an important role in the quality of groundwater, the relationship of water regulation and purification to biodiversity is poorly understood.

In lowland Europe, several factors impinge on water regulation and purification, including use of floodplains, river engineering and increasing urbanisation, leading to higher levels of run-off and contamination of water. Increasing land-use intensity and the replacement of biodiverse natural and semi-natural ecosystems by intensively managed lands and urban areas have resulted in increased run-off rates, especially in mountainous regions. Increasingly, freshwater supplies are a problem in the Mediterranean region and in such densely populated areas as southeast England.

A more coherent approach to the managed recharge of groundwater, with controls on groundwater extraction rates to protect surface ecosystems, would be a valuable enhancement to the Water Framework Directive. Trans-boundary approaches to catchment management are needed that offer a balance between engineered and ecosystem-based approaches to water regulation.

B4 Protection from Hazards

This regulating service reduces the impacts of natural forces on human settlements and the managed environment. It is highly valued in Europe. Many hazards arising in Europe from human interaction with the natural environment are sensitive to environmental change. These include flash floods due to extreme rainfall events on heavily managed ecosystems that cannot retain rainwater; landslides and avalanches on deforested slopes; storm surges, exacerbated by sea-level rise and the increasing use of hard coastal margins; air pollution due to intensive use of fossil fuels combined with extreme summer temperatures; and fires caused by prolonged drought, with or without human intervention.

Ecosystem integrity is important in protection from these hazards, but less so to geological hazards, such as volcanic eruptions and earthquakes, which are localised to a few vulnerable areas. In alpine regions, vegetation diversity is related to the risk of avalanches.[42] Soil biodiversity may play a role in flood and erosion control through affecting surface roughness and porosity,[43] and increasing tree diversity is believed to enhance protection against rockfall.[44] Increased urbanisation and more intensive use of land for production may reduce the ability of ecosystems to mitigate extreme events.

Environmental Quality Regulation

Environmental quality regulation is a new category, not in the MA. In addition to services like water purification mentioned above, ecosystems contribute to several environmental regulation services of importance for human well-being and health. Examples include the role of vegetation and green areas in urban landscapes for air cleaning, where parks may reduce air pollution by up to 85% and significantly contribute to the reduction of noise. For cities, particularly in southern Europe around the Mediterranean, vegetation and green areas may play a very important role in mitigating the urban heat island effect, a considerable health issue in view of projected climate change. Urban development in Europe, just as elsewhere in the world, faces considerable challenges where efforts to reach some environmental goals, for example, increased transport and energy efficiency through increased infilling of open space with urban infrastructure, is not done through sacrificing all other environmental qualities linked to those spaces.

B5 Pollination
The pollination service provided by ecosystems is the use of natural pollinators for crops. The role of pollinators, such as bees, in maintaining crop production is well documented and of high importance, in Europe as elsewhere in the world. There is strong evidence that loss of pollinators reduces crop yield and that the availability of a diverse pool of pollinators tends to lead to greater yields.

Habitat destruction and deterioration, with increased use of pesticides, has decreased abundance and diversity of many insect pollinators, leading to crop loss with severe economic consequences, and to reduced fecundity of plants, including rare and endangered wild species. Reduction of landscape diversity and increase of land-use intensity may lead to a reduction of pollination service in agricultural landscapes.[45,46] The loss of natural and semi-natural habitat can reduce crop production through reduced pollination services provided by native insects, including bees.[47] There is increasing evidence that the diversity of pollinators, not just abundance, may influence the quality of pollination service.[48] Maintenance of biodiverse landscapes, as well as protecting pollinators by reducing the level of use of agrochemicals (including pesticides), is an important means for sustaining pollinator service in Europe.

The concern at a European level is that change in land use, in particular urbanisation and intensive agriculture, has decreased pollination services through the loss of pollinator species. However, we do not fully understand the causes behind recent declines in pollinators.

C Provisioning Services
Provisioning services are the benefits obtained from the supply of food and other resources from ecosystems.

C1 Provision of Food
The delivery and maintenance of the food chain on which human societies depend is clearly of fundamental importance. It is estimated that well over 6000 species of plants are known to have been cultivated at some time or another,[49] but about 30 crop species provide 95% of the world's food energy.[50]

Intensive agriculture, as currently practised in Europe, is centred around crop monoculture, with minimisation of associated species. These systems offer high yields of single products, but depend on high rates of use of fertilisers and pesticides, raising questions about sustainability, both economically and environmentally. Introducing a broader range of species into agriculture might contribute significantly to improved health and nutrition, livelihoods, household food security and ecological sustainability.[51]

Maintenance of high productivity over time in monocultures almost invariably requires heavy inputs of chemicals, energy and capital, and these are unlikely to be sustainable in the face of disturbance, disease, soil erosion, overuse of natural capital (for example, water) and trade-offs with other ecosystem services.[52] Diversity may become increasingly important as a management goal, from economic and ecological perspectives, for providing a broader array of ecosystem services.

C3 Energy resources
The supply of plants for fuels represents an important provisioning service on a global scale. In Europe, traditional dependence on fuel from plants has diminished in line with the uptake of fossil fuels. However, energy from plants is expected to become more important in Europe in the future as pressures build to increase the proportion of renewable energy.

Biodiversity of the crop will probably play a small direct role in most biofuel production systems, although all land-based biofuel production will rely on the supporting and regulating services for which biodiversity is important. At present, the increase of biofuels is being achieved partly by the cultivation of biomass crops, which are burned as fuels in conventional power stations, and partly by diversion of materials otherwise useable as food for people. The expectation is that these 'first generation' fuels will be displaced – at least for ethanol production – by a second generation of non-food materials.

All of these biofuel production systems, however, present serious sustainability issues. There are already established damaging impacts on food production, availability and prices worldwide. In addition, full analyses of the carbon fluxes show that the carbon mitigation benefits are much smaller than anticipated because of losses of carbon from newly cultivated soils; destruction of vegetation when new land is brought under the plough; losses of other greenhouse gases such as nitrous oxide from nitrogen-fertilised biofuel production systems; and transport and manufacturing emissions.

Land-based biofuel production systems also have the potential to be especially damaging to conservation of biodiversity, because their introduction on a large scale will inevitably lead both to more intensive land use and to the conversion of currently uncultivated land to production. However, with the correct regulation and institutions, currently degraded land could simultaneously generate biofuels and a suite of other services as well. A full audit of the implications of increased biomass and bioenergy production is urgently needed.

C4 Provision of Fibres

The provision of fibre has historically been a highly important ecosystem service to Europe but most textiles consumed in the EU are now produced and manufactured abroad. However, the pulp and paper industry is significant in Europe and is the dominant user of plant fibres in Europe. Most raw pulp is produced from highly managed monocultures of fast-growing pine and eucalypts, grown at high densities with limited scope for biodiversity. Such large-scale monocultures are vulnerable to runaway pathogen attack.[53] Biodiverse cropping systems may prove of value for ensuring robust future productivity. Wool production is generally a low-intensity activity on semi-managed pasture lands with the potential to support considerable biodiversity.

C5 Biochemical Resources

Ecosystems provide biochemicals – materials derived from nature as feedstocks in transformation to medicines – but also other chemicals of high value such as metabolites, pharmaceuticals, nutraceuticals, crop protection chemicals, cosmetics and other natural products for industrial use. A report from the US Environmental Protection Agency[54] concludes that economically competitive products (compared with oil-derived products) are within reach, such as for celluloses, proteins, polylactides, plant oil-based plastics and polyhydroxyalkanoates. The high-value products may make use of biomass economically viable, which could become a significant land-use issue. Biodiversity is the fundamental resource for bioprospecting but it is rarely possible to predict which species or ecosystem will become an important source.[55] Harvesting for biochemicals, however, might itself have a negative impact on biodiversity if over-harvesting removes a high proportion of the species.

C6 Genetic Resources

Genetic resource provision, for example, provision of genes and genetic material for animal and plant breeding and for biotechnology, is a function of the current level of biodiversity. EU extinction rates remain low; however, there may be problems in poorly studied systems (for example, soils, marine environments). Genebanks are better developed in EU than elsewhere but have limited capacity to conserve the range of genetic diversity within populations. There are now numerous initiatives to collect, conserve, study and manage genetic resources *in situ* (for example, growing crops) and *ex situ* (for example, seed and DNA banks) worldwide, including most EU countries. New techniques, using molecular markers, are providing new precision in characterising biodiversity.[56]

D Cultural Services

Although the MA recognises many services under this heading, we have considered them in two main groups:

1. Spiritual, religious, aesthetic, inspirational and sense of place; and
2. Recreational, ecotourism, cultural heritage and educational.

All the services within these groups have a large element of non-use value, especially those in the first group to which economic value is hard to apply. Those in the second group are more amenable to traditional valuation approaches. Biodiversity plays an important role in fostering a sense of place in all European societies and thus may have considerable intrinsic cultural value.

Evidence for the importance of these services to citizens of the EU can be found in the scale of membership of conservation-oriented organisations. In the UK, for example, the Royal Society for the Protection of Birds has a membership of over one million and an annual income of over £50 million and the National Trust is even larger: 3.6 million members and an annual turnover of over £400 million. Cultural services based on biodiversity are most strongly associated with less intensively managed areas, where semi-natural biotopes dominate. These large areas may provide both tranquil environments and a sense of wilderness. Low-input agricultural systems are also likely to support cultural services, with many local traditions based on the management of land and its associated biological resources. Policy (including agricultural and forestry policies) needs to be aimed at developing sustainable land-use practices across the EU, to deliver cultural, provisioning and regulatory services effectively and with minimal cost. Maintenance of diverse ecosystems for cultural reasons can allow provision of a wide range of other services without economic intervention.

In Europe, cultural services are of critical importance because of the high value many of Europe's people place on the existence and opportunity to enjoy landscapes and open spaces with their flora and fauna. Although the intrinsic biodiversity of natural space in Europe varies greatly, there is evidence that people value 'pristine' environments and regard the impoverishment of landscape, flora and fauna as negative factors, impacting heavily on their enjoyment of nature. The economic value of ecosystems for tourism and recreation often exceeds their value for provisioning services.

The results of the assessment are summarised in Table 1; a number of ecosystems services have high importance for Europe and of these, biodiversity is important in a significant number of cases.

4 Managing Ecosystem Services in Europe

4.1 How Ecosystems Respond to Change

All ecosystems experience environmental change and disturbance, but they also have the ability to maintain themselves in the face of change. The successive appearance of distinct communities of plants and animals on a site, ecological succession, has been much studied and an important distinction between primary and secondary succession has emerged. Primary succession occurs on bare or recently uncovered surfaces such as muds, glacial moraines and river gravels. Secondary succession is the replacement of an existing community after removal of all or part of the vegetation. The major difference between the two processes is that soil has to be formed in primary succession, a process that may

Table 1   Expert opinion on the role of biodiversity in maintaining current ecosystem services in Europe.

	Increasing role of biodiversity	
Increasing importance of ecosystem service	A3: Water cycling	A1: Primary production
	A4: Soil formation	A2: Nutrient cycling
	B1: Climate regulation	B5 Pollination
	B3/C2:Water regulation and provision	D2: Cultural services: recreation
	B4: Protection from hazard	
	C1 Food provision Environmental quality	
	C3: Energy provision	B2: Disease regulation
	C4 Fibre production	C5 Biochemicals provision
	D1: Cultural services: spiritual	C6: Genetic resources

take thousands of years. Secondary succession, for example, the return of woodland to abandoned agricultural fields, depends on the ability of species to survive or disperse back into the disturbed area. If the disturbance is on a very large scale, recovery of the ecosystem can be slow.[57]

The concept of succession implies that communities recover in predictable ways after disturbance. However, species previously found on a site may fail to re-colonise. If the disturbance is on a very large scale, in space or time, the species may be extinct in the area and unable to disperse back in; for long-lived species, the local environment may have changed so much that they are no longer able to reproduce or grow from seed, either due to physical changes (*e.g.* climate change) or biotic changes (*e.g.* invasive species or a parasite). If the change is sufficiently severe, the community may shift to a new stable state, as happened in the well-documented example of the Newfoundland cod fishery, where the serious disturbance of gross and sustained over-fishing drove the population below a level from which it has been able to recover.[58]

Sustaining desirable states of an ecosystem in the face of multiple or repeated disturbance therefore requires persistence of functional groups of species.[59] Consequently, high levels of biodiversity in an ecosystem can be viewed as an insurance against major disturbance and the likelihood that the community will fail to recover to its original state, simply by increasing the chance that key species will survive or be present. The insurance metaphor can help us understand how to sustain ecosystem capacity to cope with and adapt to change, even in more complex ecosystems that have numerous possible stable states and in human-dominated environments.[60–62] In biodiverse ecosystems, species within functional groups will show a variety of responses to environmental change, and this diversity of response may be critical to ecosystem resilience. However, high species diversity does not necessarily entail high ecosystem resilience or *vice versa*, and species-rich areas may also be highly vulnerable to environmental change.

One large challenge for ecology is to predict the likely changes in ecosystems after disturbance or environmental change. Modelling tools allow improved regional estimates, and are an increasingly reliable source for estimates of ecosystem response to environmental change. As a significant example of an estimate of European ecosystem response, climate change combined with the effects of increased atmospheric CO_2 concentrations on vegetation growth were shown to produce changes in the cycling of carbon in terrestrial ecosystems.[63] Impacts were predicted to vary across Europe, showing that regional-scale studies are needed.

4.2 Threats to Biodiversity, and Consequences for Ecosystem Services in the European Union

The landscapes of Europe have altered substantially in the past 60 years, under the twin pressures of the intensification of agriculture and urbanisation. Intensive agriculture threatens delivery of many ecosystem services, especially in the European lowlands (for example, the Netherlands, parts of southern England and northern France) and in large-scale irrigation systems (for example, in Greece). The amount of carbon stored as soil organic matter has declined in most intensive arable soils and this trend is likely to continue;[64] improved management practices that take carbon sequestration as a goal could double the amount stored, with demonstrable impacts on carbon emission targets.[65] Many other examples have been documented, including threats to pollinators leading to a decline in the service of pollination;[66] increased pest problems due to the more rapid spread of pathogens through ecosystems with low biodiversity; and the impact of atmospheric nitrogen deposition on semi-natural ecosystems resulting in declines in biodiversity and poorer water quality.[67] The evidence for the effects of nitrogen deposition is clear: the long-running (more than 150 years) Park Grass experiment at Rothamsted Research in Hertfordshire, UK, shows that a species-rich grassland can be converted to a monoculture of a single grass by sustained addition of high levels of ammonium nitrogen.[68] Similarly, the almost complete loss of heathland from the Netherlands has been ascribed to atmospheric nitrogen deposition.[69]

The direct outcome of these pressures on biodiversity shows in indicators based on birds, butterflies and plants that suggest a decline of species populations in nearly all habitats in Europe: largest in farmlands, where species populations declined by an average of 23% between 1970 and 2000.[70] Large declines in agricultural landscapes of populations of pollinating insects, such as bees and butterflies, and birds, which disperse seeds and control pests, may have consequences not only on agricultural production but also on maintaining species diversity in natural and semi-natural habitats across Europe.

Urban environments have many distinctive features, the most prominent of which is their extreme heterogeneity: there are patches where both biodiversity and ecosystem service delivery is minimal, for example, where land surfaces are covered with concrete or tarmac, and others where biodiversity may be very

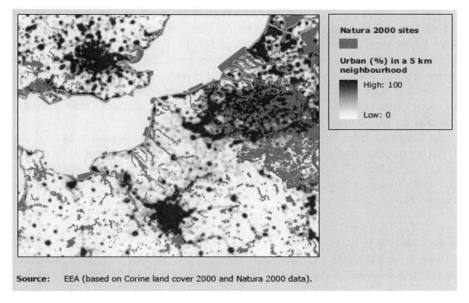

Source: EEA (based on Corine land cover 2000 and Natura 2000 data).

Figure 2   Central Belgium is composed principally of highly urbanised areas and areas of high conservation value (Natura 2000 areas). (Reproduced with kind permission of European Environment Agency.)

high, as in some gardens and parks. A consequence of this heterogeneity is the fragmentation of habitats, which favours species that are effective dispersers but militates against others. This pronounced selection leads to distinctive communities, often dominated by alien species, which by definition are good at dispersing or being dispersed. In some regions, such as central Belgium, the effect of urbanisation has been to produce a dichotomy between highly urbanised and protected areas (Figure 2).

4.3   Methods of Valuing Biodiversity and Ecosystem Services

Many threats to ecosystem services arise because of the way in which different uses of land are valued. The immediate value taken into account in decisions is typically expressed in terms of the market price of the land to a developer or the value of a crop it will produce. These approaches ignore the value of the ecosystem services provided by the land, which will be placed in jeopardy by the proposed development. The valuation of ecosystem services offers the potential to place a value on the services forfeited by the development to balance the value of the development itself in assessments of costs and benefits of alternatives. Approaches of this kind have been used widely in project evaluation both of alternative land use and for conservation investments.

The EU has taken an active role in advancing valuations through the recent TEEB (The Economics of Ecosystem Services and Biodiversity) initiative.[71] This highlights the importance of valuation of ecosystem services and the biodiversity that underpins them, and gives powerful global examples. The scoping study concludes that there are major threats to ecosystem services from the current high rate of loss of biodiversity, but that there is an emerging range of policy instruments, based on valuing ecosystem services, that provides options for managing them in future.

At the most basic level, the services provided by an ecosystem at risk can form a powerful part of the narrative in project assessment. Simply by setting down the nature of the services and their potential scale, it is possible to alter the terms of assessment so that the 'development gain' is not the only factor for consideration.

4.3.1 Quantitative Methods

In recent years, there has been considerable progress in attaching monetary value to ecosystem services and, in certain cases, to the biodiversity under-pinning them. Ecosystems have value in terms of their use, for example, for the production of food or management of flood risk. However, they also have a set of non-use values associated, for example, with the cultural and aesthetic sig-nificance they have. In many cases it has proved possible to capture both main kinds of value through a range of instruments including:

- *Revealed preference methods* based on evidence of current values as shown, (for example, in the market price of products, the impact of services on productivity or the costs associated with recreational use of landscape);
- *Cost-based methods* based on costs of replacement or damage avoided; and
- *Stated preference methods* that assess the amount people say they would be prepared to pay for ecosystem services.

Each method has strengths and weaknesses but stated preference methods, especially in the form of contingent valuation, have been most widely used in dealing with the real case of multiple services from an ecosystem. This bias reflects both an ability to handle multiple services better than the more objective methods that tend to focus on single attributes (for example, food production or flood defence) and the poor availability of the economic data that those methods require.

There is also much current interest in the development of markets for eco-system services, as exemplified by carbon trading schemes. A new tool, payment for ecosystem services (PES)[72] defines a payment for an ecosystem service as a voluntary transaction where a well-defined ecosystem service is bought by at least one buyer from at least one supplier, but only if the supplier secures the provision of the service. The transaction should be voluntary and the payment should be conditional on the service being delivered. Paying for an ecosystem service is not necessarily the same as trading nature on a market: markets may play a role, but because many ecosystem services are public goods, we cannot rely on markets alone. Actions by governments and intergovernmental organisations are also needed.

There are numerous challenges to the implementation of PES.[73] However, we lack international institutions to broker deals between suppliers of ecosystem services and the rest of the world, though some non-governmental organisations play that role for specific projects and the Global Environmental Facility (GEF), funded by all countries, is designed to deal with global conservation issues.

4.3.2   Qualitative Methods: Multi-Criteria Analysis

Generally, economic valuation of biodiversity offers ways to compare tangible benefits and costs associated with ecosystems,[74] but ignores the information about non-economic criteria (for example, cultural values) that defines biodiversity values. However, decision-making processes require knowledge of all influencing factors.[75] Multi-criteria analysis (MCA) is a structured approach for ranking alternative options that allow the attainment of defined objectives or the implementation of policy goals. A wide range of qualitative impact categories and criteria are measured according to quantitative analysis, namely scoring, ranking and weighting. The outcomes of both monetary and non-monetary objectives are compared and ranked, so that MCA facilitates the decision-making process while offering a reasonable strategy selection in terms of critical criteria.

The basis of all valuation methods, however, is an assessment of the nature and scale of the ecosystem services themselves and, in cases where the viability of the ecosystem is placed at risk, the nature and scale of the consequent impacts on the provision of ecosystem services. Where the ecosystem services are dependent on biodiversity, loss of biodiversity can be valued in terms of ecosystem services foregone or reduced, provided that there is a robust description of the relationship between biodiversity and ecosystem services. The quality of the underlying science is therefore of great significance in all kinds of valuations.

4.3.3   Putting Valuation into Practice

Methods for valuing ecosystem services and biodiversity are becoming accepted and embedded in a wide range of policy instruments.

The UK Department for Environment, Food and Rural Affairs (Defra) appraisal of options for a Flood and Coastal Erosion Risk Management (FCERM) scheme includes specific estimates of the economic value of changes in ecosystem services under a range of options, using the 'impact pathway approach'. This involves a series of steps so that a policy change; the consequent impacts on ecosystems; changes in ecosystem services; impacts on human welfare; and economic value of changes in ecosystem services are considered in turn.[76]

Key stakeholders in FCERM are broadly supportive of moves towards greater inclusion of economic value estimates in appraisals, despite the remaining uncertainty about the absolute value of the ecosystem services, resulting from uncertainty about both the physical changes in ecosystem services and the appropriate monetary values to apply to these. The authors suggest that practical appraisals 'need to compare the relative magnitude of changes in the provision of ecosystem services across different options' and conclude that this can be possible even with 'limited availability and precision of scientific and economic information. In most cases it should be possible to present a robust assessment, with suitable sensitivity analysis, highlighting the key uncertainties and exploring their implications'.[77]

The prime current example of PES, carbon trading, is developing rapidly. In Europe, the EU Emissions Trading Scheme (EU ETS) is in a second phase of development and now accounts for about 65% of global carbon trading. Current allowance prices for carbon within the EU ETS show some volatility but are currently (October 2009) around €15 (per tonne CO_2 equivalent). Volumes traded average about 8.5 million tonnes per month.

The results of valuation are increasingly recognised and accepted in policy debates and in individual decisions, on environmental impacts of projects of economic development, for example. Current knowledge of ecosystem services and the processes behind them gives a strong basis for valuation. However, it is clear that more is needed to strengthen the underpinning science.

4.4 Prioritising Ecosystem Services in Land Management: Weighing up Alternative Land Uses

There have been numerous attempts to find optimal habitat management strategies for particular broad ecosystem types, aiming to maintain biodiversity. In natural ecosystems such as forests, minimal intervention is usually the best habitat management strategy, although different types of sustainable forestry may work as well.[78] In natural aquatic ecosystems, the management of nutrient status of ecosystems is of primary importance,[79] whereas regulation of hydrology is an important issue when managing wetland ecosystems. Optimal habitat management in agricultural ecosystems requires the regulation of land-use intensity.[80] There has been much attention on semi-natural grasslands: optimal grazing and mowing regimes, techniques of cutting shrubs and burning, *etc.* have been discussed.[81] However, in all these cases the linkage to delivery of ecosystem services has been weak.

At the same time, there is accumulating evidence of the impact of land-use type and intensity on ecosystem services. For instance, the significance of European semi-natural grasslands as a source of clean and sustainably produced fodder has been recently recognised.[82] Those grasslands are extremely rich in species, but also rich in genetic variability within species and may thus provide genetic resources, which might contribute to the development of new breeds of agricultural plants, medical plants, *etc.* They also provide different regulatory services like pollination,[83] or hazard prevention,[84] or multiple cultural services. The availability of those services is primarily dependent on the continuation of the extensive land use in agricultural landscapes.

Although agri-environmental schemes encourage farmers to restore species-rich grasslands on arable land or on culturally improved pastures, the land-use types that maximise ecosystem services are not targeted in the current policies of the EU. The Common Agricultural Policy aims to increase agricultural production, without valuing ecosystem services. Similar policies apply to land use in forest or wetland ecosystems. Current policies also lack a landscape perspective and fail to take into account the linkages between landscape units or the delivery of multiple services from ecosystems. The opportunity for maintaining both ecosystem services and biodiversity outside conservation areas lies in promoting diversity of land use at the landscape and farm rather than field scale.[85] To achieve that goal, however, would require an economic and policy climate that favours diversification in land uses and diversity among land users.

Current strategies of habitat management and land use in Europe, focusing on economic benefit on the one hand and on the conservation of habitats and species of special interest on the other, now need to be broadened in order to cover a wider range of societal needs. There is therefore an urgent need for policies that prioritise the delivery of ecosystem services from land and that favour appropriate land use, encouraging habitat management and aiming to preserve or improve multiple ecosystem services. Proper ecosystem management strategies have to offer principles for land use in order to minimise the possible conflict between management goals that target different services. Besides traditionally accepted cultural services and more utilitarian services like production of food, fibre and fuel, supporting and regulatory services deserve much more attention than they have received until now.[86]

5   Conclusions

This assessment shows that the services provided to humanity by ecosystems in Europe are many, varied, of immense value and frequently not open to substitution by any artificial process. Although in some cases, biodiversity appears to play a relatively small role in maintaining ecosystem services, there is clear evidence that in others the biodiversity plays a fundamental role in the delivery of the service.

We have highlighted four of these services as being both of key importance to our survival as a society and particularly susceptible to the biological richness of the ecosystems that deliver them: primary production, nutrient cycling, pollination and a set of cultural services centred around ecotourism and recreation. Other services for which the evidence suggests that biodiversity is important appear to play a smaller role in sustaining modern European societies, at least at present.

Focussing on these services, however, may obscure a more fundamental point: that all ecosystems deliver a broad range of services, for some of which biodiversity is crucial and some of which are of particular economic or social value. Two key points arise from understanding that all ecosystems deliver multiple services:

1. Managing an ecosystem primarily to deliver one service will almost certainly reduce its ability to provide others: a forest managed exclusively for timber production will have minimal amenity and ecotouristic value, will store little carbon and will be ineffective at retaining nutrients;
2. Many of the multiple services that arise from a single ecosystem are either undervalued or completely unvalued: in the case of the forest, society currently places no value on nutrient cycling, only rarely values water cycling and regulation, and is only beginning to find ways to value carbon storage effectively.

Generally speaking, we undervalue all ecosystem services that do not provide goods that can be handled through conventional market mechanisms. Some value is placed on amenity, because of the increasing recognition that the economy of many rural areas in agriculturally marginal zones is heavily dependent on tourism, but usually the potential beneficiaries are not the managers of the land, leading to a beneficiary trade-off. No effective values are placed on most of the basic supporting services (soil formation, water and nutrient cycling) and primary production is generally only valued insofar as it creates marketable goods. Regulating services are almost always undervalued, perhaps most notably in the case of pollination, despite the fact that in this case it is possible to understand the value that it provides in relation to marketable goods such as food.

There is an urgent need therefore to provide incentives to managers of land and water to ensure the maintenance of the broad range of services from the ecosystems that they manage. Because of the difficulty of using traditional economic instruments to achieve this goal, an alternative regulatory framework is needed, which may require the development of a set of binding legal requirements, as in the case of the Water Framework Directive.

The research that has been assessed in this chapter demonstrates that both the quality and quantity of biodiversity are important for maintaining the health of ecosystems and their ability to deliver services to society. The importance of biodiversity varies greatly among services, being particularly strong for primary production, nutrient cycling and pollination, for example, but much less so for protection from natural hazards. The way in which biodiversity ensures the processes that underly ecosystem services is only partly understood, and there is an urgent need for research to determine how great a loss of biodiversity can be experienced before service delivery declines.

References

1. P. M. Vitousek, H. A. Mooney, J. Lubchenco and J. M. Melillo, *Science*, 1997, **277**, 494.

2. IUCN, *Redlist of Threatened Species*, 2004.
3. S. Rojstaczer, S. M. Sterling and N. J. Moore, *Science*, 2001, **294**, 2549.
4. R. B. Jackson, S. R. Carpenter, C. N. Dahm, D. M. McKnight, R. J. Naiman, S. L. Postel and S. W. Running, *Ecol. Applicat.*, 2001, **11**, 1027.
5. D. Pearce, in *Frontiers in Biodiversity Economics*, ed. A. U. Kontoleon, U. Pascual and T. Swanson, Cambridge University Press, Cambridge, UK, 2006.
6. A. H. Fitter, T. Elmqvist, A. Rinaldo, H. Setälä, S. Stoll-Kleemann and M. Zobel, *Ecosystem Services and Biodiversity in Europe*, EASAC, London, 2009.
7. A. G. Tansley, *Forest Ecol. Manag.*, 1935, **16**, 284.
8. C. Folke, S. Carpenter, B. Walker, M. Scheffer, T. Elmqvist, L. Gunderson and C. S. Holling, *Ann. Rev. Ecol. Evolut. Systematics*, 2004, **35**, 557.
9. P. B. Reich, D. Tilman, S. Naeem, D. S. Ellsworth, J. Knops, J. Crane, D. Wedin and J. Trost, *Proc. Natl. Acad Sci. U. S. A.2004*, **101**, *10101*.
10. J. A. Estes and D. O. Duggins, *Ecol. Monogr.*, 1995, **65**, 75.
11. J. Terborgh, L. Lopez, P. Nunez, M. Rao, G. Shahabuddin, G. Orihuela, M. Riveros, R. Ascanio, G. H. Adler, T. D. Lambert and L. Balbas, *Science*, 2001, **294**, 1923.
12. P. M. Vitousek and L. R. Walker, *Ecol. Monogr.*, 1989, **59**, 247.
13. D. U. Hooper, F. S. Chapin, L. L. Ewel, A. Hector, P. Inchausti, S. Lavorel, J. H. Lawton, D. M. Lodge, M. Loreau, S. Naeem, B. Schmid, H. Sctälä, A. J. Symstad, J. Vandermeer and A. Wardle, *Ecol. Monogr.*, 2005, **75**, 3.
14. C. Folke, S. Carpenter, B. Walker, M. Scheffer, T. Elmqvist, L. Gunderson and C. S. Holling, *Ann. Rev. Ecol. Evolut. Systematics, 2004*, **35**, *557*.
15. P. Balvanera, A. B. Pfistera, N. Buchmann, J.-S. He, T. Nakashizuka, D. Raffaelli and B. Schmid, *Ecol. Lett.*, 2006, **9**, 1146.
16. P. Hoehn, T. Tscharntke, J. M. Tylianakis and I. Steffan-Dewenter, *Proc. R. Soc. London, Ser. B*, 2008, **275**, 2283.
17. J. M. Bullock, R. F. Pywell and K. J. Walker, *J. Appl. Ecol.*, 2007, **44**, 6.
18. G. E. Likens, F. H. Bormann, N. M. Johnson, D. W. Fisher and R. S. Pierce, *Ecol. Monogr.*, 1970, **40**, 23.
19. P. Smith, *Soil Use Manag.*, 2004, **20**, 212.
20. T. Elmqvist, C. Alfsen and J. Colding, in *Encyclopedia of Ecology*, ed. S. E. Jorgensen and B. D. Fath, Elsevier, Oxford, UK, 2008, **vol. 5**.
21. P. Bolund and S. Hunhammar, *Ecol. Econ.*, 1999, **29**, 293.
22. G. S. Lovasi, J. W. Quinn, K. M. Neckerman, M. S. Perzanowski and A. Rundleet, *J. Epidemiol. Community Health*, 2008. doi:10.1136/jech.2007.071894.
23. F. Worrall, A. Armstrong and J. K. Adamson, *J. Hydrol.*, 2007, **339**, 1.
24. A. H. Fitter, T. Elmqvist, A. Rinaldo, H. Setälä, S. Stoll-Kleemann and M. Zobel, *Ecosystem Services and Biodiversity in Europe*, EASAC, London, 2009.
25. P. H. Ciais, M. Reichstein, N. Viovy, A. Grainer, J. Ogée, V. Allard, M. Aubinet, N. Buchmann, Chr. Bernhofer, A. Carrara, F. Chevallier, N. de

Noblet, A. D. Friend, P. Friedlingstein, T. Grünwald, B. Heinesch, P. Keronen, A. Knohl, G. Krinner, D. Loustau, G. Manea, G. Matteucci, F. Miglietta, J. M. Oureival, D. Papale, K. Pilegaard, S. Rambal, G. Seufert, J. F. Soussana, M. J. Sanz, E. D. Schutze, T. Vesala and R. Valentini, *Nature*, 2005, **437**, 529.

26. Q. Schiermeier, *Nature*, 2008, **455**, 572.
27. E. A. B. Eltahir and R. L. Bras, *Rev. Geophys.*, 1996, **34**, 367.
28. T. P. Barnett, J. C. Adam and D. P. Lettenmaier, *Nature*, 2005, **438**, 303.
29. I. A. Rodriguez-Iturbe, A. Porporato, L. Ridolfi, V. Isham and D. R. Cox, *Proc. R. Soc. London, Ser. A*, 1999, **455**, 3789.
30. A. Darracq, F. Greffe, F. Hannerz, G. Destouni and V. Cvetkovic, *Water Sci. Technol.*, 2005, **51**, 31.
31. F. Giorgi, P. Whetton, R. Jones, J. H. Christensen, L. O. Mearns, B. Hewitson and H. Von Storch, *Geophys. Res. Lett.*, 2001, **28**, 3317.
32. E. I. Newman, *J. Appl. Ecol.*, 1997, **34**, 1334.
33. J. W. A. Poesen and J. M. Hooke, *Progr. Phys. Geogr.*, 1997, **21**, 157.
34. I. A. Janssens, A. Freibauer, P. Ciais, P. Smith, G. J. Nabuurs, G. Folberth, B. Schlamadinger, R. W. A. Hutjes, R. Ceulemans, E.-D. Schulze, R. Valentini and A. J. Dolman, *Science*, 2003, **300**, 1538.
35. H. H. Janzen, *Agri. Ecosyst. Environ.*, 2004, **104**, 399.
36. P. H. Bellamy, P. J. Loveland, R. I. Bradley, R. M. Lark and G. J. D. Kirk, *Nature*, 2005, **437**, 245.
37. M. Kulmala, T. Suni, K. E. J. Lehtinen, M. Dal Maso, M. Boy, A. Reissell, U. Rannik, P. Aalto, P. Keronen, H. Hakola, J. Back, T. Hoffmann, T. Vesala and P. Hari, *Atmos. Chem. Phys.*, 2004, **4**, 557.
38. D. H. Wall and R. A. Virginia, in *Nature and Human Society: The Quest for a Sustainable World*, ed. P. Raven and T. A. Williams, National Academy of Sciences Press, 2000, p. 225.
39. S. Sánchez-Moreno and H. Ferris, *J. Nematol.*, 2006, **38**, 290.
40. W. Zhang, T. Ricketts, C. Kremen, K. Carney and S. Swinton, *Ecol. Econ.*, 2007, **64**, 253.
41. R. Naylor and P. Ehrlich, in *Nature's Services: Societal Dependence on Natural Ecosystems*, ed. G. Daily, Island Press, Washington, D.C, 1997, p.151.
42. F. Quetier, S. Lavorel, W. Thuiller and I. Davies, *Ecol. Appl.*, 2007, **17**, 2377.
43. P. Lavelle, T. Decaëns, M. Aubert, S. Barot, M. Blouin, F. Bureau, P. Margerie, P. Mora and J. P. Rossi, *Eur. J. Soil Biol.*, 2006, **42**, 3.
44. L. K. A. Dorren, F. Bergera, A. C. Imesonb, B. Maiere and F. Reya, *Forest Ecol. Manag.*, 2004, **195**, 165.
45. T. Tscharntke, A. M. Klein, A. Kruess, I. Steffan-Dewenter and C. Thies, *Ecol. Lett.*, 2005, **8**, 857.
46. E. Öckinger and H. G. Smith, *J. Appl. Ecol.*, 2007, **44**, 50.
47. T. H. Ricketts, J. Regatz, I. Steffan-Dewenter, S. A. Cunningham, C. Kremen, A. Bogdanski, B. Gemmil-Herren, S. S. Greenleaf, A. M. Klein, M. M. Mayfield, L. A. Morandin, A. Ochieng and B. F. Viana, *Ecol. Lett.*, 2008, **11**, 499.

48. P. Hoehn, T. Tscharntke, J. M. Tylianakis and I. Steffan-Dewenter, *Proc. R. Soc. London, Ser. B*, 2008, **275**, 2283.
49. V. H. Heywood, *Use and Potential of Wild Plants in Farm Households*, FAO, Rome, 1999.
50. J. Williams and N. Haq, *Global Research on Underutilized Crops: An Assessment of Current Activities and Proposals for Enhanced Cooperation*, ICUC, Southampton, UK, 2002.
51. H. Jaenicke and I. Höschle-Zeledon, *Strategic Framework for Underutilized Plant Species Research and Development, with Special Reference to Asia and the Pacific and to Sub-Saharan Africa*, International Centre for Under-utilised Crops Colombo Sri Lanka and Global Facilitation Unit for Underutilized Species, Rome, Italy, 2006.
52. D. U. Hooper and F. S. Chapin III, *Ecol. Monogr.*, 2005, **75**, 3.
53. K. E. Mock, B. J. Bentz, E. M. O'Neill, J. P. Chong, J. Orwin and M. E. Pfrende, *Mol. Ecol.*, 2007, **16**, 553.
54. US Environmental Protection Agency, *Bioengineering for Pollution Prevention through Development of Biobased Materials and Energy*, State-of-the-Science Report. US EPA Washington, DC, 2007.
55. A. J. Beattie, W. Barthlott, E. Elisabetsky, R. Farrel, C. T. -Kheng, I. Prance, J. Rosenthal, D. Simpson, R. Leakey, M. Wolfson and K. ten Kate, in *Ecosystems and Human Well-Being, Volume 1, Current State and Trends; Millennium Ecosystem Assessment*, ed. R. Hassan, R. Scholes and N. Ash, Island Press, Washington, 2005, p. 273.
56. R. Fears, *Genomic and Genetic Resources for Food and Agriculture, Background Study Paper No. 34*, Commission for Genetic Resources for Food and Agriculture, FAO, Rome (published on FAO website), 2007.
57. P. M. Attiwill, *Forest Ecol. Manag.*, 1994, **63**, 247.
58. R. A. Myers, J. A. Hutchings and N. J. Barrowman, *Ecol. Appl.*, 1997, **7**, 91.
59. J. Lundberg and F. Moberg, *Ecosystems*, 2003, **6**, 87.
60. C. Folke, C. S. Holling and C. Perrings, *Ecol. Appl.*, 1996, **6**, 1018.
61. J. Norberg, D. P. Swaney, J. Dushoff, J. Lin, R. Casagrandi and S. A. Levin, *Proc. Natl. Acad. Sci. U. S. A.*, 2001, **98**, 11376.
62. G. W. Luck, G. C. Daily and P. R. Ehrlich, *Trends Ecol. Evol.*, 2003, **18**, 331.
63. P. Morales, T. Hickler, D. P. Rowell, B. Smith and M. T. Sykes, *Global Change Biol.*, 2006, **13**, 108.
64. J. Smith, P. Smith, M. Wattenbach, S. Zaehle, R. Hiederer, R. J. A Jones, L. Montanarella, M. D. A. Rounsevell, I. Reginster and F. Ewert, *Global Change Biol.*, 2005, **11**, 2141.
65. K. Paustian, J. Six, E. T. Elliott and H. W. Hunt, *Biogeochemistry*, 2000, **48**, 147.
66. T. Tscharntke, A. M. Klein, A. Kruess, I. Steffan-Dewenter and C. Thies, *Ecol. Lett.*, 2005, **8**, 857.
67. G. K. Phoenix, W. K. Hicks, S. Cinderby, J. C. I. Kuylenstierna, W. D. Stock, F. J. Dentener, K. E. Giller, A. T. Austin, R. D. B. Lefroy, B. S. Gimeno, M. R. Ashmore and P. Ineson, *Global Change Biol.*, 2006, **12**, 470.

68. M. Dodd, J. Silvertown, K. McConway, J. Potts and M. Crawley, *J. Ecol.*, 1995, **83**, 277.
69. J. Bobbink, M. Hornung and J. G. M. Roelofs, *J. Ecol.*, 1998, **86**, 717.
70. M. de Heer, V. Kapos and B. J. E. ten Brink, *Philos. Trans. R. Soc. London, Ser. B*, 2005, **360**, 297.
71. http://ec.europa.eu/environment/nature/biodiversity/economics/pdf/teeb_report.pdf.
72. S. Wunder, *Payments for Environmental Services: Some Nuts and Bolts*, CIFOR, Occasional Paper No. 42 Wunder, 2005.
73. A. H. Fitter, T. Elmqvist, A. Rinaldo, H. Setälä, S. Stoll-Kleemann and M. Zobel, *Ecosystem Services and Biodiversity in Europe,* EASAC, London, 2009.
74. S. Pagiola, K. von Ritter and J. Bishop, *Assessing the Economic Value of Ecosystem Conservation,* The World Bank Environment Department, Washington DC, 2004.
75. OECD, *Recommendation of the Council on the Use of Economic Instruments in Promoting the Conservation and Sustainable Use of Biodiversity; Endorsed by Environment Ministers, adopted by the OECD Council*, 2004, vol. 2006.
76. Defra, *An Introductory Guide to Valuing Ecosystem Services*, Wildlife-Countryside Section, Defra, London, 2007, p. 22.
77. Defra, *An Introductory Guide to Valuing Ecosystem Services*, Wildlife-Countryside Section, Defra, London, 2007, p. 49.
78. T. Kuuluvainen, *Silva Fennica*, 2002, **36**, 5.
79. A. Baattrup-Pedersen, S. E. Larsen and T. Riis, *Hydrobiologia*, 2002, **495**, 171.
80. M. D. A. Rounsevell, P. M. Berry and P. A. Harrison, *Environ. Sci. Policy*, 2006, **9**, 93.
81. P. Poschlod and M. F. Wallis de Vries, *Biological Conservation*, 2002, **104**, 361.
82. J. M. Bullock, R. F. Pywell and K. J. Walker, *J. Appl. Ecol.*, 2007, **44**, 6.
83. T. Tscharntke, A. M. Klein, A. Kruess, I. Steffan-Dewenter and C. Thies, *Ecol. Lett.*, 2005, **8**, 857.
84. F. Quetier, S. Lavorel, W. Thuiller and I. Davies, *Ecol. Appl.*, 2007, **17**, 2377.
85. M. J. Swift, A.-M. N. Izac and M. van Noordwijk, *Agric. Ecosyst. Environ.*, 2004, **104**, 113.
86. E. Nelson, G. Mendoza, J. Regetz, S. Polasky, H. Tallis, D. R. Cameron, K. Chan, G. C. Daily, J. Goldstein, P. M. Kareiva, E. Lonsdorf, R. Naidoo, T. H. Ricketts and M. R. Shaw, *Front. Ecol. Environ.*, 2009, **7**, 4.

Ecosystem Services and Policy: A Review of Coastal Wetland Ecosystem Services and an Efficiency-Based Framework for Implementing the Ecosystem Approach

PIRAN C. L. WHITE, JASMIN A. GODBOLD, MARTIN SOLAN, JESSICA WIEGAND AND ALISON R. HOLT

ABSTRACT

The Ecosystem Approach (EA) to environmental management aims to enhance human well-being within a linked social and ecological system, through protecting the delivery of benefits and services to society from ecosystems in the face of external pressures such as climate change. However, our lack of understanding of the linkages between the human and natural components of ecosystems inhibits the implementation of the EA for policy decision-making. Coastal wetland systems provide many benefits and ecosystem services to humans, including nutrient recycling, climate and water quality regulation, timber, fuel and fibre, but they are under considerable threat from population pressure and climate change. In this chapter, we review the ecosystem services provided by coastal wetlands, and the threats to these services. We then present a new integrative conceptual framework to underpin the EA. The framework is divided into three sub-systems: one relating to ecosystem functions, one to ecosystem services, and one to social development and well-being. The pathways linking these sub-systems represent transfers of state, for example, ecosystem functions being transferred into ecosystem services, or ecosystem services being transferred into benefits. The focus of our approach is on enhancing the *magnitude* and *efficiency* of these transfers,

Issues in Environmental Science and Technology, 30
Ecosystem Services
Edited by R.E. Hester and R.M. Harrison
© Royal Society of Chemistry 2010
Published by the Royal Society of Chemistry, www.rsc.org

by introducing or making use of any existing *catalysts* and overcoming any *constraints* in the system. The framework represents a dynamic system for implementing the EA in which interventions can be planned and managed in an adaptive way.

1   Ecosystem Services and the Ecosystem Approach to Policy

Erosion of global ecological resources,[1] combined with a growing evidence of the impacts of anthropogenic environmental change,[2–4] have focused attention on how land, air and water resources can be managed in a more adaptive manner to produce continuing benefits to society.[5–7] Approaches based on the conservation of particular land- or sea-based habitats are limited as they only protect or sustain specific resources and are spatially constrained, covering just 12% of the Earth's terrestrial surface[8] and considerably less of the Earth's marine environment (0.72%).[9] Ensuring that humans continue to derive benefits from the environment requires a multifunctional approach, which takes specific consideration of the need to maintain the health and resilience of ecological resources into the future.

The Ecosystem Approach (EA), as articulated in the Millennium Ecosystem Assessment,[10] is based on the principles of sustainable development with the twin goals of increasing both human and ecosystem well-being.[11] Humans are central to the EA and related approaches, and the benefits that humans derive from the environment are viewed in terms of ecosystem goods and services, which may be regulating, provisioning, supporting or cultural.[10] The aim of the EA is to enhance social utility or well-being within a healthy and resilient ecosystem, which is able to maintain its delivery of ecosystem services in the face of various human-induced pressures internal to the system, as well as external pressures, such as extreme events in the short term or climate change in the longer term.

In recent years the EA has increasingly featured in the environmental policies of various countries, including the US[12,13] and the UK.[14] For the first time, environmental policy development is making explicit links between biodiversity conservation and socio-economic development,[12] thus recognising the interdependence of enhancing social well-being and conserving threatened ecosystems. This represents a paradigm shift in policy terms but one that poses significant challenges for both policy-makers and the scientific research community,[15] since there are no clear frameworks for guiding the application of the EA and evaluating its success.

In this chapter, we review the ecosystem services provided by coastal wetlands, which are under considerable threat from global environmental change. We then consider some of the policy challenges posed in the management of coastal wetlands. Finally, we introduce a new conceptual framework based on notions of sustainability and efficiency, which can be used to implement the ecosystem approach in conserving ecosystem services and enhancing social well-being.

2 Existing Frameworks for Understanding Ecosystem Services

There is no single agreed definition of ecosystem services. However, a number of definitions have been developed over the last decade. Daily[16] proposed that ecosystem services were 'conditions and processes through which natural ecosystems sustain and fulfil human life'. Thus, they served a life support function and provided 'ecosystem goods' such as food, foliage and timber, which can be harvested to enhance economic and social well-being. This classification has since been refined by De Groot *et al.*[17] who defined 'goods' and' 'services' together as a product of 'ecosystem functions', which in turn emerged from 'structures' and 'processes'. In their definition, 'processes' are the result of interactions between biotic (living organisms) and abiotic (chemical and physical) components of ecosystems through the universal driving forces of matter and energy. 'Functions' represent the result of these processes, and these produce 'goods' and 'services' that satisfy human needs, directly or indirectly. With minor variations, these broad classifications have been retained by subsequent authors.[10,18–20]

In recent years, there has been much emphasis on the economic valuation of ecosystem services. For example, Costanza *et al.*[21] estimated the value of global ecosystem services at \$33 $268 \times 10^9 \, yr^{-1}$. This paper caused much controversy among environmental economists since it was based on estimations of aggregated 'total economic value', which are inconsistent with the marginal approach underpinning economic cost-benefit analysis.[22,23] Because of the difficulty of valuing certain ecosystem services and specifically to avoid problems with double-counting in ecosystem service valuation,[23] Fisher *et al.*[24] distinguished between direct and indirect ecosystem services. In their framework, ecosystem services are those aspects of ecosystems utilised actively or passively to produce human well-being. Ecosystem functions result from processes occurring in the ecosystem as a result of the interactions between biodiversity and the physico-chemical environment. Ecosystem functions become services if humans gain direct or indirect benefits from them, but without humans there are no ecosystem services. Thus, in this framework, nutrient cycling is a process that results in clean water. If the clean water is consumed, then the clean water is a benefit of the directly utilised service of clean water provision, which results from the indirectly utilised service of nutrient cycling.

One of the characteristics of all these frameworks is that they are inherently linear and represent a production chain. Thus, physical and biological processes combine to produce functions and services which are consumed by humans. However, one of the central tenets of the Ecosystem Approach is that humans are an integral part of complex ecosystems (social-ecological systems)[25] within which there are likely to be important feedback loops. Thus, consumption of ecosystem services by humans at the end of the chain has implications for future production of these services through its impacts on biological and physical processes at the start of the chain. The presence of feedbacks is recognised by the authors of the above frameworks,[20] but the frameworks themselves do not enable these to be considered in any detail. Furthermore, the focus of these frameworks on the natural science components of the system means that their value for guiding decisions on sustainable natural resource management is severely limited. As yet our knowledge of both the systems themselves, and especially the inter-linkages between the natural and social components, is frequently lacking. This represents a major problem for policy-makers.

3 Coastal Wetlands: Ecosystems on the Front Line of Global Change

Estuarine and coastal areas have been the focal point for human settlement and marine resource use throughout history:[26,27] today approximately 40% of the world's population lives within 100 km of the coast, which is more than three times higher than the global average density.[28,29] Coastal areas are subjected to intense human pressures (see Table 1a)[27] because a significant proportion of the human population depend on these areas for food, shelter, economic prosperity and well-being. Human populations in coastal areas have traditionally had close links with the sea, whether in terms of direct benefits such as fishing or trade, or indirect ones such as recreation. In addition to the human pressures, coastal wetlands are exposed to a wide variety of natural threats (*e.g.* droughts, floods and other climatic extremes, see Table 1a).[30] A large risk is posed by climate change because these habitats are frequently low-lying and therefore particularly susceptible to the effects of sea level rise, such as increased tidal inundation and coastal erosion. Coastal wetlands are frequently also at the forefront of strategic responses to mitigate the impacts of sea level rise, such as coastal realignment.[23] As well as their direct economic importance, many coastal areas are vital for biodiversity conservation and the maintenance of ecosystem services, such as nutrient cycling, climate regulation and water regulation. Yet, many coastal areas also suffer from various social problems, such as poor quality housing, and high levels of social deprivation[31] and economic regeneration is hampered by their physical isolation and poor transport links. Thus, in most coastal wetland ecosystems the physico-chemical environment, biodiversity and human society are inextricably linked. Against this backdrop of environmental, social and political pressures, implementation of an eco-system approach to enhance sustainable development and human well-being in the face of climate change faces significant challenges.

4 Defining Coastal Wetlands

The delineation of marine-associated habitats is less developed than equivalent classification schemes in terrestrial systems, and is complicated by ambiguous and shared descriptions of particular components of the habitat that transcend the terrestrial-marine interface. A lack of baseline descriptions and rudimentary data on the spatial extent of individual areas that persist over time as a recognisable community, and the absence of a common terminology within a single comprehensive and authoritative classification scheme, complicates habitat identification and subsequent assessment of status.[32] Wetland habitats are a case in point as, in general, they share the characteristics of both terrestrial and aquatic habitats because they occupy the transitional zones between permanently wet and generally dry environments. Nevertheless, the Ramsar Convention on Wetlands of International Importance (Article 1.1) formally defines wetlands as 'areas of marsh, fen, peatland or water, whether natural or artificial, permanent or temporary, with water that is static or flowing, fresh, brackish or salt, including areas of marine water, the depth of which at low tide does not exceed six metres', and (Article 2.1) which 'may incorporate riparian and coastal zones adjacent to the wetlands, and islands or bodies of marine water deeper than six metres at low tide lying within wetlands', thereby including rivers and shallow coastal waters (http://www.ramsarg.org/). Thus coastal wetlands specifically include several ecosystem types, including deltas and estuaries, tidal flats, seagrass beds, salt marshes, saline lagoons, mangroves and coral reefs, which are distinguished largely based on their bio-physical habitat characteristics.[33]

Coastal wetland habitats occur globally and cover most latitudes (see Figure 2.2 in ref. 10), although it is difficult to quantify the total coverage of specific types of coastal wetlands because individual habitats are not always distinguished when area estimates are made. However, it has been estimated that approximately $500\,000\,km^2$ of coastal area is covered by estuaries, and tidal flats are thought to extend to $300\,000\,km^2$ worldwide,[33,34] but these estimates do not account for recent and substantial habitat losses in estuarine areas (including tidal flats and salt marsh). In California, for example, less than 10% of natural coastal wetlands remain, and in other countries many estuarine areas (including tidal flats and salt marsh) have been substantially altered or entirely lost as a result of land reclamation.[34,35] On the Essex coastline of the UK, approximately 40 000 ha of saltmarsh have been lost following the construction of medieval to 19th century embankments.[36] Seagrasses cover approximately $165\,000\,km^2$ worldwide, although estimates have varied up to $600\,000\,km^2$ (ref. 37,38). Seagrass beds are predominantly found in inlets and lagoons that are sheltered from strong wave action. Within the UK there are three different species of seagrass (*Zostera* sp.) all of which are considered to be scarce, with the largest continuous populations covering approximately 1200 ha in the Cromarty Firth (Scotland) and 325 ha at Maplin Sands (England).[39] Globally, an estimated $12\,000\,km^2$ of seagrass meadows were lost in the 1990s, corresponding to an area of about 2%,[40] largely as a result of direct human impacts, such as dredging, fishing, eutrophication, aquaculture, as well as indirect impacts due to climate change (*e.g.* sea level rise and increased storm events).[37] Mangroves are largely a feature of sheltered, tropical coastlines and cover approximately $200\,000\,km^2$ (ref. 30,41), but are disappearing at a global rate of 1 to 2%. In a large number of countries, mangroves have been cleared at a rate of 50–80% over the past 15 years, largely as a result of shrimp aquaculture development, deforestation, freshwater diversion, pollution and upstream land use.[30,42,43]

Table 1 Summary of (a) the principal anthropogenic and natural threats to coastal wetlands, (b) the ecosystem functions and (c) ecosystem services and benefits provided by coastal wetland habitats.

	Estuary	Saltmarsh	Mangrove	Tidal flat	Seagrass	Coral reef	Saline Lagoon
a) Threats							
Climate change	x	x	x	x	x	x	
Sea level rise	x	x	x	x		x	x
Flood defences/barrage schemes	x	x		x			x
Physical disturbance/habitat destruction	x						x
Land claim		x	x	x	x	x	x
Erosion		x	x			x	
Sedimentation		x			x	x	
Over exploitation	x		x	x		x	
Pollution	x		x		x	x	x
Eutrophication	x	x	x	x	x	x	
Invasive species	x	x	x	x	x	x	
Grazing		x				x	
Hypoxia	x					x	
Acidification	x	x	x	x	x	x	x
Ice reduction	x	x					x
Disease				x	x	x	x

	1	2	3	4	5	6	7
b) *Ecosystem functions*							
Nutrient cycling	X	X	X	X	X		X
Carbon sequestration	X		X	X	X		
Decomposition	X	X	X	X	X		
Primary production	X	X	X	X	X	X	X
Carbon cycling	X	X	X	X	X	X	
c) *Ecosystem services and benefits*							
Biological control and regulation	X	X	X	X	X		X
Freshwater storage and retention	X	X		X			
Hydrological Balance	X	X		X			
Atmospheric and climate regulation		X	X	X	X		
Human disease control	X	X	X	X	X		
Waste processing	X	X	X	X	X		
Flood/storm protection	X	X	X	X	X		X
Erosion control and sediment retention	X	X	X	X	X		
Cultural and amenity	X	X	X	X	X		
Recreational	X	X	X	X	X		X
Aesthetics	X	X	X	X	X		X
Education and research	X	X	X	X	X		X
Biochemical	X	X	X	X	X		X
Nutrient cycling and fertility	X	X	X	X	X		X
Biodiversity	X	X	X	X	X		X
Food	X	X	X	X	X		X
Fibre, Timber, Fuel	X	X	X		X		X

Thus, similar to most marine habitats,[26,44–46] changes in biodiversity and coastal wetland ecosystems are caused directly by exploitation, pollution and habitat destruction, or indirectly through climate change and related perturbations of ocean biochemistry (see Table 1a). Coastal wetlands are greatly threatened by climate change, specifically sea level rise. The Intergovernmental Panel on Climate Change (IPCC) projections estimate that rising atmospheric temperatures have increased the global heat content in the upper 300 m of the oceans at a rate of about 0.04 °C decade^{-1}.[47] As a result of thermal expansion of water, global sea level rise predictions are estimated to reach up to 82 cm by 2100 (ref. 48), resulting in extensive areas of coastal wetlands lost to coastal squeeze. Such physical changes will have significant implications for the future distribution of coastal wetland habitats, their fauna and flora, and associated ecosystem processes.[49] Although it has recently been suggested that a warming-induced stimulation of saltmarsh growth in northern latitudes is comparable to the estimates of tidal marsh area that will be lost due to sea level rise,[50] in general it is unlikely that coastal areas will accrete at a rate that exceeds water level rise; indeed, accretion will have to occur at a rate two to seven times that observed over the last century just to match projected rises within the next century.[51] Similarly, even in delta areas where accretion rates are much greater than anticipated rises in sea level, alteration of river discharge patterns and sediment loads by human activities may significantly reduce natural levels of resilience.[52] Thus, although the ecosystem services of certain coastal wetland habitats may be maintained under anticipated future climate change scenarios, a less optimistic outcome is more realistic when taking a global view.

5   Ecosystem Services from Coastal Wetlands

Human societies have been built on biodiversity, as they have learnt to use the diversity of organisms directly for medicines, food and fibres, and it is now also well-established that biodiversity has strong effects on a number of ecosystem services (*e.g.* waste processing and flood protection) by mediating ecosystem processes and functions, such as nutrient cycling and primary production (see Table 1b–c; ref. 53,54). A wealth of empirical and theoretical studies has shown that biodiversity loss has largely a negative effect on ecosystem functioning,[55] although research on the contribution of biodiversity to ecosystem services is still in its infancy[56] (but see also ref. 57,58). Coastal wetlands provide vital ecosystem services, such as water quality and climate regulation, are valuable accumulation sites for sediment, contaminants, carbon and nutrients, and very importantly offer protection from coastal erosion and storm surges. In addition, they provide vital nursery and breeding grounds for birds, fish, shellfish, crustaceans and mammals, as well as renewable resources such as timber, fuel and fibre.[10,43,59] Based on calculations by Costanza *et al.*[21] the global value of coastal wetlands (including estuaries, seagrass, coral reefs, tidal marsh and mangroves) is around one third ($9934 \times 10^9 \, yr^{-1}$) of the total global value across 16 terrestrial and marine biomes. Thus coastal wetlands are a valuable ecological and economic resource, yet they are increasingly degraded as a result of human activities, which may lead to the long-term loss or changes in the delivery of ecosystem services provided by these habitats. Indeed the importance of ecosystems and the services they provide is often only recognized after they have been lost (*e.g.* Hurricane Katrina).[60]

Coastal wetland ecosystem services occur at multiple scales, from climate regulation and carbon sequestration at the global scale, to flood protection, water supply, nutrient cycling and waste treatment at the local and regional scales. In addition, ecosystems do not have sharp boundaries, but rather overlap and/or interact at multiple spatial and temporal scales.[61] Thus for effective management of ecosystem services to occur, the spatial and temporal scales over which ecosystem dynamics, management issues and societal impacts occur have to be identified, and scales must be consistent with the ability to recognize and explain the most important drivers and threats to the ecosystem.[62] The efficiency of ecosystems to provide ecosystem services, however, is highly variable in space and time and not all coastal wetlands perform all services equally well.[58,63]

The ecosystem or physical spatial area of ecosystem service provision is the scale at which the joint processes work to provide the ecosystem functions on which the desired service(s) depend. However, the social spatial scale of ecosystem services is the scale at which humans benefit from the desired outcomes, which will vary between services.[64] Flood protection by saltmarshes and mangroves will occur not only in adjacent areas, but will also be felt further upstream and away from the actual wetland area. For example, a mangrove forest of 1 km width can protect communities up to 5 km inland from tropical storms.[65] However, the scale at which land management and conservation strategies of ecosystem services occur, also depend on the outcome preference of society and the preference of the stakeholder(s) for which the wetland is managed. Barbier *et al.*[65] highlighted the complexities involved in future coastal management and the importance of considering and examining multiple ecosystem services provided by coastal wetlands. In this example it was shown that conversion of mangroves to shrimp ponds, although directly beneficial to the shrimp farmer and outside investors, would have detrimental effects on the region's coastal ecosystem service provision, including flood protection, wood resources and nursery habitat for commercially important fish species. This suggests that the economic gains that can occur from habitat conversion may be outweighed by the potential benefits of habitat conservation, especially as multiple ecosystem services are provided by an ecosystem.[66] Thus, in order for the service-providing units to effectively map onto the management units, land management strategies have to incorporate the ecosystem services society wants from coastal wetlands and then determine how these can be best realised, whilst taking into consideration what services coastal wetlands have the potential to provide sustainably.

6 Management to Combat Environmental Change and Threats to Coastal Wetlands

A significant number of anthropogenic threats to marine ecosystems have been documented (see Table 1a) and are known to negatively affect at least 85% of the coastline[67] and most, if not all, of the global ocean.[27] These include climate-related effects, such as sea level rise, increased storm frequency, rising temperature and elevated carbon dioxide (CO_2) levels,[68] as well as direct human intervention through the overexploitation of resources[45] or coastal development. Irrespective of the cause of change, recent trends in the global status of major biomes broadly indicate high rates of conversion from natural habitat to modified landscapes that are structurally less complex and less able to support prior levels of biodiversity or maintain ecosystem functioning and services,[66] although the short- and long-term susceptibility of individual species and habitats to drivers of change can vary considerably with physiographic setting.[69] The loss of wetlands in some of the world's major deltas provides an indication of the extent of direct human influence, where the rate of habitat conversion (average across all major deltas: $95 \, km^2 \, year^{-1}$; range: $1–419 \, km^2 \, year^{-1}$) has resulted in a 52.4% reduction in total delta plain area.[70] Other connected human activities, such as the inland construction of dams to secure drinking water, has substantially increased the likelihood of further habitat deterioration in the delta areas by reducing rates of sediment replenishment.[71] These rates and levels of conversion are not exceptional and are typical of many other wetland habitats (reviewed in ref. 72), despite the expectation that the human requirement for ecosystem services from these regions is likely to exceed provisioning capacity in the near future.[73]

There is a clear challenge in managing resources and adapting to ecosystem changes when potential benefits to human well-being are being eroded at the present pace and at this order of magnitude. The ecological consequences of anthropogenic forcing can lead to dramatic changes in species composition and functioning (*e.g.* ref. 45). This may include non-linear and accelerating ecosystem responses,[65,74] especially when the additive, synergistic or antagonistic affects of multiple drivers are considered.[75,76] Whilst an emerging literature base recognises that the combined effects of multiple stressors on ecosystem functioning can diverge from those predicted from the effects of single elements (*e.g.* biodiversity × environmental heterogeneity;[77] biodiversity × CO_2 × temperature[78]), incorporating these complex interactions and feedbacks into a single management framework raises difficulties. This is because any subsequent decision-making process must be context dependent and will be subject to multiple, often conflicting, societal needs and wishes. In this decision context it is necessary to determine the appropriate spatial scale over which desirable ecosystem services are distributed;[79–81] identify any relevant interactions across scales (*e.g.* spatial scale *versus* jurisdiction);[82] establish the temporal variance in service supply and demand; ascertain the relative role of the abiotic and biotic components of the environment in determining levels of service;[83] and determine socio-economic and cultural requirements, whilst accounting for drivers of change that modify both the environment and human dimensions.

The process of balancing these competing demands can be informed by understanding how a suite of key ecological variables and processes relate to the package of ecosystem services society requests,[84] before emphasising the level at which peoples needs will be met, in both the short and longer term.[24] It is naïve to suggest that individual ecosystem services can be substituted or replaced using technology, as single service substitutions ignore the multi-functionality, resilience and interconnectedness of ecosystems, and the methodologies used to maintain such efforts inevitably affect other components of natural systems.[85] Attempts to supplement ecological services with technological solutions have been valuable, however, as the reasons for failure have served to highlight the complexities and dynamics of natural systems.[86] Hence, a more fruitful approach is to build on existing knowledge in an attempt to establish how the components that underpin service delivery and social context interrelate and affect the provision and utilisation of a range of ecosystem services.[87] Whilst this approach is intuitive, and a cursory consideration of the literature would suggest that socio-ecologists are well placed to extract the relevant information, the fragmentary nature of the detail hinders assembly of the necessary integrated overview. For example, whilst the notion that biological diversity positively regulates ecosystem functioning is generally accepted[88] and supported by rigorous theoretical, empirical and observational studies (for reviews see ref. 55,89,90), how biodiversity–ecosystem functioning relations translate to ecosystem services is less clear, despite a detailed understanding of society's dependence on natural systems.[10] The assumption that biodiversity regulates ecosystem functioning, and that individual functions in turn underpin particular services that are regarded as important for human well-being, suggests a linear process that, although conceptually intuitive, argues that the conservation of biodiversity alone increases the likelihood of service provision.[54] Indeed, Luck *et al.*[79] have argued that at the local level, species are the fundamental unit underpinning service delivery because the level of service provision will be proportional to the functional traits of the organisms that deliver services.[56,91] This view, however, ignores the contribution of other components of the ecosystem and is built on a repository of selected ecosystem services (see Table 2 in ref. 56), such as biological control, pollination and seed dispersal, where the biological contribution of individual species equates directly to the service. Where multiple components of the ecosystem, including indirectly associated biotic and abiotic variables, are considered as potential contributors to ecosystem functioning and service provision, they tend to be integral to the process under study.[77,78] Indeed, under certain circumstances, abiotic variables can be more important than biodiversity *per se*, especially within perturbed systems[83] and/or at larger scales.[92]

A systematic way to bind multiple (abiotic×biotic) components of the ecosystem with socio-economic information, whilst being inclusive to stakeholders' opinions and wishes, is to engage in a participatory process that seeks to determine a fair and transparent decision-making process. A participatory approach has a number of potential benefits, such as encouraging social learning about the complexities of the system being managed;[93] helping to identify alternative values and solutions; increasing fairness in decision making; and reducing conflict.[94,95] It can also enhance social capacity by providing opportunities for the acquisition of new skills by local people, providing a platform for coordination between organisations, and creating a sense of local ownership and responsibility.[96] It is essential that this is an adaptive process, to account for the emergence of further information about the complexities of the system over time and as human demands change. Such participation will create the social conditions necessary to achieve adaptive governance,[97,98] essential for adaptive ecosystem management. For salmon fisheries in the North Pacific, for example, fisheries management regimes traditionally assumed that stocks would remain stable provided certain climatic criteria were maintained, a view that was naïve to regime shifts in the population associated with non-linear responses to climatic fluctuations.[86] Whilst native human populations were able to adapt to such changes,[99] a socio-ecological structure surrounding modern day salmon populations now features prominently in the processes that lead to adopted management practices and coping strategies.[100]

However, the difficulty with this approach is that when established levels of human well-being are at risk, the choice for individuals is often to reduce their standard of living or accept any associated costs with preserving species and ecosystems that underpin service provision.[101] The conundrum here is that preserving the latter is necessary for the long-term viability of the former, but sustainable resolutions to short-term conflicts can often only be achieved by enforced regulation, as compliance may have short-term negative consequences. A topical example is that of the herring fisheries off New England, where the estimated biomass of the fish population is above maximal sustainable yield for supporting a fishery, but establishment of a fishery may cause a regime shift that will have cascading effects on the entire ecosystem, potentially resulting in the collapse of several fisheries. Clearly, achieving a long-term sustainable fishery whilst maintaining a resilient ecosystem will be at the expense of economically significant short-term gain.[102] It is important to note, however, that the decision process can only be effective if the adaptive capacity of local communities are in line with the restrictions imposed, otherwise any socio-ecological problems are only likely to be exacerbated.[103] In the longer term, it would seem prudent to actively pursue a change in societal values to support mechanisms that respond to a range of non-economic metrics relevant to ecosystem services. This process will need to demonstrate the value (social, cultural, economic *etc.*) and benefit of maintaining or restoring ecological resources at sustainable levels, whilst communicating levels of uncertainty and reassuring local communities across the span of likely socio-ecological futures.[104] In order to achieve this, it is clear that the preservation of service provision requires the amalgamation of beneficiaries, service providers, and the abiotic and biotic components of ecosystems that provide them,[56] with equivalent socio-economic and cultural trends, the modifying effects of environmental change on these, and the overarching governance structures, into a unifying conceptual framework to underpin and inform subsequent ecosystem-based management decisions.[105]

7 A New Conceptual Framework to Underpin the Ecosystem Approach

One of the principal barriers to establishing such ecosystem-based operational frameworks in the past has been the non-commensurability of the units which can be used to quantify ecosystem functions, services, goods and benefits. However, the notion of sustainability, with its emphasis on efficiency, provides a means by which we can establish a common measurement unit by focusing not on the functions, services or goods themselves, but on the processes of conversion between them. In this section, we present a unifying framework to underpin the ecosystem approach, based around magnitudes and efficiencies of conversion (Figure 1).

The framework is divided into three sub-systems: one broadly relating to ecosystem functions; one to ecosystem services; and one to social development and well-being. However, these sub-systems are linked, and there are feedback mechanisms occurring both within and between these sub-systems. Each pathway in the system represents a transfer of state, for example, ecosystem functions being transferred into ecosystem services, and ecosystem services being transferred into benefits. The focus of this approach is on the *magnitude* and *efficiency* of these transfers, whether in terms of biodiversity interactions with the environment delivering ecosystem functions; ecosystem functions being translated into ecosystem services through human use; or benefits from ecosystems being converted into social well-being. The spatial and temporal dynamics of transfers are also important, since very different time scales may operate in different parts of the system, both in terms of the rate of transfers of state, but also in relation to management interventions.

The key to enhancing sustainability, and social utility or well-being, is to understand the opportunities and barriers in the system, whether these are ecological, social or economic. These can be considered as *catalysts* and *constraints* in the system, such as in a chemical reaction. This requires an understanding of the transfer relationships, including potential environmental/social limits or tipping points (*e.g.* ecological meltdown or social inertia) and non-linearity; the constraints around them, *e.g.* the social or environmental context; and also their spatial and temporal dynamics. In some circumstances, scientific uncertainty may also act as a constraint to developing efficiency; progress towards sustainability can be made by reducing uncertainty through improved understanding of the interdependencies of particular parts of the system. The framework represents a dynamic system in which interventions can be potentially managed in an adaptive way.

In the past, integrated social-economic-ecological frameworks have been limited by the non-commensurability of units between different components of the system. The novelty of this framework is threefold:

1. The focus for evaluation is on the magnitude and efficiencies of transfer and how these are affected by (or are a function of) the components

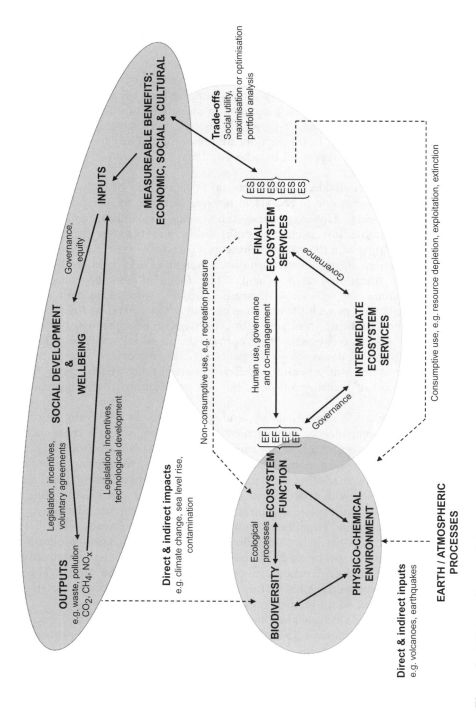

Figure 1 Efficiency framework for an ecosystem services approach to sustainability.

contributing to them. This permits the inclusion of environmental limits and thresholds where these occur and are understood.

2. The framework has the capacity to be truly adaptive, based on dynamic sustainability functions built up from real time data of social and environmental processes. This enables the incorporation of natural fluctuations, limits and uncertainty into the sustainability functions.

3. The framework is both integrative and inter-disciplinary and is not constrained by units of measurement. Efficiencies can be measured in different ways depending on the parts of the system with which we are concerned, and the metrics can be entirely quantitative, or a combination of qualitative and quantitative.

The framework provides a conceptual basis for assessing the components of social-ecological systems and the links between them, which is vital for implementing the ecosystem approach and informing sustainable development and economic growth. We can consider sustainability in terms of maximising the magnitude and/or efficiency of transfers subject to various limits in the system. This can be considered in terms of ecological or socio-economic sustainability, and could be formally quantified in terms of optimizing certain transfers, or by maximizing them subject to certain constraints, such as thresholds. This then allows us to incorporate notions of resilience and vulnerability in ecological systems, the adaptability of societies in terms of political structures and governance, and to evaluate the extent to which policy is fit for purpose in terms of the enhancing sustainability.

As an example, magnitude and/or efficiency of nutrient cycling in estuarine benthic systems as an ecosystem function can be set in purely quantitative terms as a production function in terms of the processes which underpin it, such as bioturbation and bioirrigation, or in terms of the species whose behaviour gives rise to these processes, potentially incorporating in addition some characteristic(s) of the physical environment (*e.g.* organic content, turbidity, current regimes). The functions can be built up as knowledge increases, and can specifically incorporate human-mediated catalysts and constraints, so the impact of these can be quantified. In the nutrient cycling example, constraints might include pollution in the short term or acidification in the longer term. In a social example, the magnitude and/or efficiency in conversion of benefits to sustainable economic growth could be measured in terms of 'sustainability' indicators such as quantitative measures of social equality,[106] or qualitative measures such as subjective assessments of well-being. Examples of catalysts or constraints could include locally-directed social-environmental development programmes in the short term, or changes in government policy in the longer term. Environmental economics has already provided examples of how changes in ecosystem service provision can be translated into monetary terms, and this could be used as another measure of efficiency at the sub-system level. In addition, social efficiency could be quantified using established social network metrics, certain network topologies indicating the effectiveness of governance.[107,108]

In terms of implementation, quantification of efficiencies at different places within the system can be used to quantify sustainability, and reductions in magnitude and/or efficiency in space or over time would represent reduced sustainability. Based on our dynamic sustainability functions, we would be able to quantify the inherent variation of sustainability measures and determine how external drivers, such as climate change, modify this variability as well as the overall trends. If the processes of transfer are poorly understood, then they can be substituted by measurements of the inputs or end products, or marginal changes in these, in common with the economic valuation approach. For example, in relation to the nutrient example, we could measure nutrient concentrations (as an output), but also measure organic matter input and phytoplankton (production) in the water column. Measuring the biomass and diversity of the processing community (*i.e.* the invertebrates) is another option. For social processes, such as assessments of well-being, we can measure the extent to which people are benefiting, but we can also measure the variables that influence that, for example, the use of green spaces or participation in community groups.

8 Conclusions and Future Challenges

Ecosystem services are increasingly at the heart of environmental policy, especially in the context of environmental change. The broader acknowledgement of the complex inter-relationships between ecological and social systems highlights the need to develop new conceptual frameworks through which the ecosystem approach can be implemented and evaluated. The efficiency transfer based framework we have introduced here provides a means by which this can be potentially achieved.

One major hindrance to the implementation of the Ecosystem Approach is the difficulty of reconciling the holistic ethos of the framework with existing property rights structures. The Ecosystem Approach requires the development of multifunctional landscapes, capable of sustaining the delivery of a range of ecosystem services. Implicit within this is an appreciation that certain trade-offs may need to be made between different land uses and the services that these provide. For some services, these decisions can be made at the property level, but for other services, trade-offs and collaboration may be needed between neighbouring properties to ensure delivery of a service at the landscape scale. Developing the necessary mechanisms to ensure the delivery of ecosystem services represents a key challenge to policy-makers. However, our framework provides a means by which trade-offs can be quantified, and integrated with a participatory process will aid decision-making.

We have focused in the discussion above on quantifying and evaluating efficiencies within specific parts of the sub-systems, but a major advance would be to develop means of comparing efficiencies between different sub-systems within the overall framework. One possible approach would be the use of techniques such as Cost Utility Analysis (CUA) and Threat Reduction

Assessment (TRA). CUA was first developed to evaluate health-care pro-grammes by comparing the outputs of competing alternatives in terms of the utility that they provide, where utility refers to the improvement in health status following a treatment, and is measured in Quality-Adjusted Life Years (QALY).[109–111] Cullen *et al.*[112] developed the Conservation Output Protection Year (COPY) to serve an equivalent function in conservation evaluation as the QALY does in health-care evaluation, but with utility now referring to the improvement in conservation status following the implementation of a con-servation programme. Within our framework, we could apply this in terms of improvements in the efficiency of ecosystem service or benefit provision. TRA was developed by Salafsky and Margoluis[113] as a technique for using progress in reducing threats to biodiversity as a proxy measurement of conservation success.[113] Threat Reduction Assessment is particularly well-suited to con-sidering the key external drivers affecting particular transfers in the system and the effectiveness of management interventions. TRA could be applied within our framework in terms of evaluating the changes in threats to specific transfer processes, such as the threats to ecosystem functions or service delivery, or the threats to the equitable delivery of benefits to society. It may also be possible to use the Threat Reduction Index to make comparisons across the sub-systems, and therefore identify areas where differences in system management have been most effective, to develop effective co-management and highlight where more needs to be done to reduce threats or improve efficiency or sustainability. One of the significant advantages of both CUA and TRA is that they require no monetary valuation, and thus overcome many of the problems inherent in the valuation of ecosystem services.[23] Finally, various other techniques from economics, industrial ecology and management, such as production function approaches,[114] life cycle assessment,[115] emergy analysis[116,117] and materials flow analysis,[118] may also be applicable to such systems-level efficiency-based ana-lysis of ecosystem services and sustainability.

If we are to conserve multiple ecosystem services at the same time as main-taining economic development and enhancing social well-being, an ecosystem-level approach to management and policy-making is essential. Management needs to be on a long-term basis, and ensuring sustainability of ecosystem services will require trade-offs that may lead to undesirable consequences for some sectors of society in the short term.[119] Nevertheless, there are increasing examples of how an ecosystem-level approach to environmental decision-making can bring substantial benefits[12,114] and can also attract greater external support and funding, *e.g.* from the private sector, than traditional habitat-or species-based approaches.[120] The Ecosystem Approach to management, with its reliance on stakeholder involvement and participatory processes, is demanding to implement, but the potential dividends across all sectors of society are substantial. We have proposed a conceptual framework through which the impacts of the competing demands of different sectors of society on ecosystem functions and services can be evaluated, and key constraints and catalysts affecting ecosystem service provision and sustainability can be iden-tified and determined. This framework can be applied in a participatory and adaptive manner at a variety of scales for the long-term benefit of society and the environment on which it depends.

Acknowledgements

This work was supported by a transdisciplinary seminar series award from the Economic and Social Research Council and the Natural Environment Research Council in the UK (Grant ref. RES-496-26-0040). We are grateful to the other participants in the seminar series for their stimulating and insightful discussions.

References

1. F. S. Chapin III, E. S. Zavaleta, V. T. Eviner, R. L. Naylor, P. M. Vitousek, H. L. Reynolds, D. U. Hooper, S. Lavorel, O. E. Sala, S. E. Hobbie, M. C. Mack and S. Diaz, *Nature*, 2000, **405**, 234–242.
2. G.-R. Walther, E. Post, P. Convey, A. Menzel, C. Oarmesan, T. J. C. Beebee, J.-M. Fromentin, O. Hoegh-Guldberg and F. Bairlein, *Nature*, 2002, **416**, 389–395.
3. C. Parmesan and G. Yohe, *Nature*, 2003, **421**, 37–42.
4. A. L. Perry, P. J. Low, J. R. Ellis and J. D. Reynolds, *Science*, 2005, **308**, 1912–1915.
5. G. Peterson, G. A. de Leo, J. J. Hellmann, M. A. Janssen, A. Kinzig, J. R. Malcolm, K. L. O'Brien, S. E. Pope, D. S. Rothman, E. Shevliakova and R. R. T. Tinch. *Conserv. Ecol.*, 1997, **1**(2), 4. http://www.consecol.org/vol1/iss2/art4/
6. E. L. Tompkins and W. Adger, *Ecol. Soc.*, 2004, **9**, 10. http://www.ecologyandsociety.org/vol9/iss2/art10/
7. P. E. Hulme, *J. Appl. Ecol.*, 2005, **42**, 784–794.
8. T. M. Lee and W. Jetz, *Future Battlegrounds for Conservation under Global Change. Proc. R. Soc. London, Ser. B*, 2008, **265**, 1261–1270.
9. M. Spalding, L. Fish and L. Wood, *Conserv. Lett.*, 2008, **1**, 217–226.
10. MEA, *Ecosystems and Human Well-being: Wetlands and Water Synthesis*, World Resources Institute, Washington DC, 2005.
11. S. M. Garcia and K. L. Cochrane, *ICES J. Marine Sci.*, 2005, **62**, 311–318.
12. H. Tallis, P. Kareiva, M. Marvier and A. Change, *Proc. Natl. Acad. Sci. U. S. A.*, 2008, **105**, 9457–9464.
13. S. R. Carpenter, H. A. Mooney, J. Agard, D. Capistrano, R. De Fries, S. Diaz, T. Dietz, A. K. Duraiappah, A. Oteng-Yeboah, H. M. Pereira, C. Perrings, W. V. Reid, J. Sarukhan, R. J. Scholes and A. Whyte, *Proc. Natl. Acad. Sci. U. S. A.*, 2009, **106**, 1305–1312.
14. Defra, *Securing a Healthy Natural Environment: An Action Plan for Embedding an Ecosystems Approach*, Defra, London, 2007.
15. M. Elliott, D. Burdon, K. L. Hemingway and S. E. Apitz, *Estuarine Coast. Shelf Sci.*, 2007, **74**, 349–366.

16. G. C. Daily, *Nature's Services: Societal Dependence on Natural Ecosystems*, Island Press, Washington DC, 1997.
17. R. A. de Groot, M. A. Wilson and R. M. J. Boumans, *Ecol. Econ.*, 2002, **41**, 393–408.
18. J. Boyd and S. Banzhaf, *Ecol Econ.*, 2006, **63**, 616–626.
19. K. Wallace, *Biol. Conserv.*, 2007, **139**, 235–246.
20. R. Haines-Young and M. Potschin, England's Terrestrial Ecosystem Services and the Need for an Ecosystem Approach, Report to Defra, Project NR0107, 2008.
21. R. Costanza, R. d'Arge, R. de Groot, S. Farber, M. Grasso, B. Hannon, K. Limburg, S. Naeem, R. V. O'Neill, J. Paruelo, R. Raskin, P. Sutton and M. van den Belt, *Nature*, 1997, **387**, 253–260.
22. M. Toman, *Biol. Conserv.*, 1998, **25**, 57–60.
23. R. K. Turner, D. Burgess, D. Hadley, E. Coombes and N. Jackson, *Global. Environ. Change*, 2007, **17**, 397–407.
24. B. Fisher, R. K. Turner and P. Morling, *Ecol. Econ.*, 2009, **68**, 643–653.
25. F. Berkes and C. Folke, *Linking Social and Ecological Systems. Management Practices and Social Mechanisms for Building Resilience*, Cambridge University Press, Cambridge, 1998.
26. H. K. Lotze, H. S. Lenihan, B. J. Bourque, R. H. Bradbury, R. G. Cooke, M. C. Kay, S. M. Kidwell, M. X. Kirby, C. H. Peterson and J. B. C. Jackson, *Science*, 2006, **312**, 1806–1809.
27. B. S. Halpern, S. Walbridge, K. A. Selkoe, C. V. Kappel, F. Micheli, C. d'Agrosa, J. F. Bruno, K. S. Casey, C. Ebert, H. E. Fox, R. Fujita, D. Heinemann, H. S. Lenihan, E. M. P. Madin, M. T. Perry, E. R. Selig, M. Spalding, R. Steneck and R. Watson, *Science*, 2008, **319**, 948–952.
28. J. E. Cohen, C. Small, A. Mellinger, J. Gallup and J. Sachs, *Science*, 1997, **278**, 1211–1212.
29. C. Small and R. J. Nicholls, *J. Coastal Res.*, 2003, **19**, 584–599.
30. S. Diop, *Estuarine Coast. Shelf Sci.*, 2003, **58**, 1–2.
31. Communities and Local Government Committee (CLGC), *Coastal Towns. Second Report of Session 2006–2007*, HC 351, Stationery Office, House of Commons, London, 2007.
32. S. Fraschetti, A. Terlizzi and F. Boero, *J. Exp. Marine Biol. Ecol.*, 2008, **366**, 109–115.
33. E. Wolanski, M. M. Brinson, D. R. Cahoon and G. M. E. Perillo, in *Coastal Wetlands: An Integrated Ecosystem Approach*, ed. G. M. E. Perillo, E. Wolanski, D. R. Cahoon and M. M. Brinson, Elsevier, Amsterdam, 2009.
34. UNEP, *Marine and Coastal Ecosystems and Human Well-Being: A Synthesis Report Based on the Findings of the Millennium Ecosystem Assessment*, UNEP, 2006.
35. E. Wolanski, *Estuarine Ecohydrology*, Elsevier, Amsterdam, 2007.
36. A. M. Dixon, D. J. Leggett and R. C. Weight, *J. Inst. Water Environ. Manag.*, 1998, **12**, 107–112.

37. C. M. Duarte, *Environ. Conserv.*, 2002, **29**, 192–206.
38. E. P. Green and F. T. Short, *World Atlas of Seagrasses,* University of California Press, Berkeley, 2003.
39. A. Maddock, *UK Biodiversity Action Plan: Priority Habitat Descriptions,* 2008. http://www.ukbap.org.uk/library/UKBAPPriorityHabitatDescriptions finalAllhabitats20081022.pdf.
40. F. T. Short and S. Wyllie-Echeverria, *Environ. Conserv.*, 1996, **23**, 17–27.
41. N. C. Duke, J. O. Meynecke, S. Dittmann, A. M. Ellison, K. Anger, U. Berger, S. Cannicci, K. Diele, K. C. Ewel, C. D. Field, N. Koedam, S. Y. Lee, C. Marchand, I. Nordhaus and F. A. Dahdouh-Guebas, *Science*, 2007, **317**, 41–42.
42. I. Valiela, J. L. Bowen and J. K. York, *Bioscience*, 2001, **51**, 807–815.
43. D. M. Alongi, *Environ. Conserv.*, 2002, **29**, 331–349.
44. J. M. Pandolfi, R. H. Bradbury, E. Sala, T. P. Hughes, K. A. Bjorndal, R. G. Cooke, D. McArdle, L. McClenachan, M. J. H. Newman, G. Paredes, R. R. Warner and J. B. C. Jackson, *Science*, 2003, **301**, 955–958.
45. J. B. C. Jackson, M. X. Kirby, W. H. Berger, K. A. Bjorndal, L. W. Botsford, B. J. Bourque, R. H. Bradbury, R. Cooke, J. Erlandson, M. A. Estes, T. P. Hughes, S. Kidwell, C. B. Lange, H. S. Lenihan, J. M. Pandolfi, C. H. Peterson, R. S. Steneck, M. J. Tegner and R. R. Warner, *Science*, 2001, **293**, 629–638.
46. B. Worm, M. Sandow, A. Oschlies, H. K. Lotze and R. A. Myers, *Science*, 2005, **309**, 1365–1369.
47. N. L. Bindoff, J. Willebrand, V. Artale, A. Cazenave, J. M. Gregory, S. Gulev, K. Hanawa, C. Le Quéré, S. Levitus, Y. Nojiri, C. K. Shum, L. D. Talley and A. S. Unnikrishnan, in *Climate Change 2007 The Physical Science Basis,* ed. S. Solomon, D. Qin, M. Manning, Z. Chen, M. Marquis, K. B. Avery, M. Tignor and H. L. Miller, Cambridge University Press, Cambridge and New York, 2007, 387–432.
48. M. Siddal, T. F. Stocker and P. U. Clark, *Nat. Geosci.*, 2009, **2**, 571–575.
49. T. Fujii, *Estuarine Coast. Shelf Sci.*, 2007, **75**, 101–119.
50. M. L. Kirwan, G. R. Guntenspergen and J. T. Morris, *Global Change Biol.*, 2009, **15**, 1982–1989.
51. J. W. Day, R. R. Christian, D. M. Boesch, A. Yáñez-Arancibia, J. Morris, R. R. Twilley, L. Naylor, L. Schaffner and C. Stevenson, *Estuarine Coast.*, 2008, **31**, 477–491.
52. J. P. M. Syvitski, C. Vörösmarty, A. J. Kettner and P. Green, *Science*, 2005, **308**, 376–380.
53. S. Diaz, J. Fargione, F. S. Chapin and D. Tilman, *PLoS Biol.*, 2006, **4**, 1300–1305.
54. S. R. Palumbi, P. A. Sandifer, J. D. Allan, M. W. Beck, D. G. Fautin, M. J. Fogarty, B. S. Halpern, L. S. Incze, J. A. Leong, E. Norse, J. J. Stachowicz and D. H. Wall, *Front. Ecol. Environ.*, 2009, **7**, 204–211.
55. P. Balvanera, A. B. Pfisterer, N. Buchmann, J. S. He, T. Nakashizuka, D. Raffaelli and B. Schmid, *Ecol. Lett.*, 2006, **9**, 1–11.

56. G. W. Luck, R. Harrington, P. A. Harrison, C. Kremen, P. M. Berry, R. Bugter, T. P. Dawson, F. de Bello, S. Diaz, C. K. Feld, J. R. Haslett, D. Hering, A. Kontogianni, S. Lavorel, M. Rounsevell, M. J. Samways, L. Sandin, J. Settele, M. T. Sykes, S. van den Hove, M. Vandewalle and M. Zobel, *Bioscience*, 2009, **59**, 223–235.
57. B. Worm, E. B. Barbier, N. Beaumont, J. E. Duffy, C. Folke, B. S. Halpern, J. B. C. Jackson, H. K. Lotze, F. Micheli, S. R. Palumbi, E. Sala, K. A. Selkoe, J. J. Stachowicz and R. Watson, *Science*, 2006, **314**, 787–790.
58. E. W. Koch, E. B. Barbier, B. R. Silliman, D. J. Reed, G. M. E. Perillo, S. D. Hacker, E. F. Granek, J. H. Primavera, N. Muthiga, S. Polasky, B. S. Halpern, C. J. Kennedy, C. V. Kappel and E. Wolanski, *Front. Ecol. Environ.*, 2009, **7**, 29–37.
59. A. P. Covich, M. C. Austen, F. Barlöcher, E. Chauvet, B. J. Cardinale, C. L. Biles, P. Inchausti, O. Dangles, M. Solan, M. O. Gessner, B. Statzner and B. Moss, *Bioscience*, 2004, **54**, 767–775.
60. J. Q. Chambers, J. I. Fisher, H. C. Zeng, E. L. Chapman, D. B. Baker and G. C. Hurtt, *Science*, 2007, **318**, 1107–1107.
61. L. Hein, K. van Koppen, R. S. de Groot and E. C. van Ierland, *Ecol. Econ.*, 2005, **57**, 209–228.
62. P. S. Levin, M. J. Fogarty, S. A. Murawski and D. Fluharty, *PLoS Biol.*, 2009, **7**, 23–28.
63. J. B. Zedler, *Front. Ecol. Environ.*, 2003, **1**, 65–82.
64. K. M. A. Chan, M. R. Shaw, D. R. Cameron, E. C. Underwood and G. C. Daily, *PLoS Biol.*, 2006, 4(11), e379. doi:10.13171/journal.pbio. 0040379nAQ2.3n.
65. E. B. Barbier, E. W. Koch, B. R. Silliman, S. D. Hacker, E. Wolanski, J. Primavera, E. F. Granek, S. Polasky, S. Aswani, L. A. Cramer, D. M. Stoms, C. J. Kennedy, D. Bael, C. V. Kappel, G. M. E. Perillo and D. J. Reed, *Science*, 2008, **319**, 321–323.
66. A. Balmford, A. Bruner, P. Cooper, R. Costanza, S. Farber, R. E. Green, M. Jenkins, P. Jefferiss, V. Jessamy, J. Madden, K. Munro, N. Myers, S. Naeem, J. Paavola, M. Rayment, S. Rosendo, J. Roughgarden, K. Trumper and R. K. Turner, *Science*, 2002, **297**, 950–953.
67. L. Airoldi and M. Beck, *Oceanogr. Marine Biol.*, 2007, **45**, 347–407.
68. A. S. Brierley and M. J. Kingsford, *Curr. Biol.*, 2009, **19**, R602–R614.
69. A. S. Kolker, S. L. Goodbred Jr, S. Hameed and J. Cochran, *Estuarine Coast. Shelf Sci.*, 2009, **84**, 493–508.
70. J. M. Coleman, O. K. Huh and B. DeWitt Jr, *J. Coastal Res.*, 2008, **24**, 1–14.
71. J. P. M. Syvitski, A. J. Kettner, I. Overeem, E. W. H. Hutton, M. T. Hannon, G. R. Brakenridge, J. Day, C. Vörösmarty, Y. Saito, L. Giosan and R. J. Nicholls, *Nat. Geosci.*, 2009, **2**, 681–686.
72. L. Airoldi, D. Balata and M. W. Beck, *J. Exp. Marine Biol. Ecol.*, 2008, **366**, 8–15.

73. J. Alcamo, D. van Vuuren, C. Ringler, W. Cramer, T. Masui, J. Alder and K. Schulze, *Ecol. Soc.* 2005, **10**, 19. http://www.ecologyandsociety.org/vol10/iss2/art19/

74. A. Dobson, D. Lodge, J. Alder, G. S. Cumming, J. Keymer, J. McGlade, H. Mooney, J. A. Rusak, O. Sala, V. Wolters, D. Wall, R. Winfree and A. Xenopoulos, *Ecology*, 2006, **87**, 1915–1924.

75. C. L. Folt, C. Y. Chen, M. V. Moore and J. Burnaford, *Limnol. Oceanogr.*, 1999, **44**, 864–877.

76. R. Przeslawski, S. Ahyong, M. Byrne, G. Wörheide and P. Hutchings, *Global Change Biol.*, 2008, **14**, 2773–2795.

77. K. E. Dyson, M. T. Bulling, M. Solan, G. Hernandez, D. Raffaelli, P. C. L. White and D. M. Paterson, *Proc. R.. Soc. London, Ser. B.*, 2007, **274**, 2547–2554.

78. M. T. Bulling, N. Hicks, L. Murray, D. M. Paterson, D. Raffaelli, P. C. L. White and M. Solan, *Philos. Trans. R. Soc. London, Ser. B*, 2009, in press.

79. G. W. Luck, G. C. Daily and P. R. Erlich, *Trends Ecol. Evol.*, 2003, **18**, 331–336.

80. G. Cognetti and F. Maltagliata, *Marine Pollut. Bull.*, 2009, **58**, 637–638.

81. C. R. Johnson, *Ecol. Soc.*, 2009, **14**, 7. http://www.ecologyandsociety.org/vol14/iss1/art7/

82. D. W. Cash, W. Adger, F. Berkes, P. Garden, L. Lebel, P. Olsson, L. Pritchard and O. Young, *Ecol. Soc.*, 2006, **11**, 8. http://www.ecologyandsociety.org/vol11/iss2/art8/

83. J. A. Godbold and M. Solan, *Marine Ecol. Prog. Ser.*, 2009, **396**, 281–290.

84. J. N. Sanchirico and P. Mumby, *Theor. Ecol.*, 2009, **2**, 67–77.

85. F. Moberg and P. Rönnbäck, *Ocean Coast. Manag.*, 2003, **46**, 27–46.

86. D. L. Bottom, K. K. Jones, C. A. Simenstad and C. L. Smith, *Ecol. Soc.* 2009, **14**, 5. http://www.ecologyandsociety.org/vol14/iss1/art5/

87. E. M. Bennett, G. D. Peterson and L. J. Gordon, *Ecol. Lett.*, 2009, **12**, 1394–1404.

88. F. Schläpfer and B. Schmid, *Ecol. Appl.*, 1999, **9**, 893–912.

89. B. J. Cardinale, D. S. Srivastava, J. E. Duffy, J. P. Wright, A. L. Downing, M. Sankaran and C. Jouseau, *Nature*, 2006, **443**, 989–992.

90. B. Schmid, P. Balvanera, B. J. Cardinale, J. Godbold, A. B. Pfisterer, D. Raffaelli, M. Solan and D. S. Srivastava, in *Biodiversity, Ecosystem Functioning and Human Wellbeing: An Ecological and Economic Perspective,* ed. S. Naeem, D. E. Bunker, A. Hector, M. Loreau and C. Perrings, Oxford University Press, Oxford, 2009.

91. C. Kremen, *Ecol. Lett.*, 2005, **8**, 468–479.

92. B. J. Anderson, P. R. Armsworth, F. Eigenbrod, C. D. Thomas, S. Gillings, A. Heinemeyer, D. B. Roy and K. J Gaston, *J. Appl. Ecol.*, 2009, **46**, 888–896.

93. C. Pahl-Wostl, M. Craps, A. Dewulf, E. Mostert, D. Tabara and T. Taillieu, *Ecol. Soc.*, 2007, **12**, 5. http://www.ecologyandsociety.org/vol12/iss2/art5/

94. L. S. Jackson, *Local Environ.*, 2001, **6**, 135–147.

95. M. S. Reed, *Biol. Conserv.*, 2008, **141**, 2417–2431.
96. N. Bracht and A. Tsouros, *Health Promot. Int.*, 1990, **5**, 199–208.
97. T. Dietz, E. Ostrom and P. C. Stern, *Science.*, 2003, **302**, 1902–12.
98. C. Folke, T. Hahn, P. Olsson and J. Norberg, *Annu. Rev. Environ. Resources*, 2005, **30**, 441–73.
99. S. J. Langdon, in *Native Americans and the Environment: Perspectives on the Ecological Indian,* ed. M. E. Harkin and D. R. Lewis, University of Nebraska Press, Lincoln, Nebraska, USA, 2007.
100. M. C. Healey, *Ecol. Soc.*, 2009, **14**, 2. http://ecologyandsociety.org/vol14/iss1/art2/
101. F. Meyerson, *Front. Ecol. Environ.*, 2009, **7**, 511.
102. A. Bakun, E. A. Babcock and C. Santorra, *ICES J. Marine Sci.*, 2009, **66**, 1768–1775.
103. C. Camargo, J. H. Maldonado, E. Alvarado, R. Moreno-Sánchez, S. Mendoza, N. Manrique, A. Mogollón, J. D. Osorio, A. Grajales and J. A. Sánchez, *Biodiversity Conserv.*, 2009, **18**, 935–956.
104. C. H. Peterson and R. N. Lipcius, *Marine Ecol. Prog. Ser.*, 2003, **264**, 297–307.
105. E. F. Granek, S. Polasky, C. V. Kappel, D. J. Reed, D. M. Stoms, E. W. Koch, C. J. Kennedy, L. A. Cramer, S. D. Hacker, E. B. Barbier, S. Aswani, M. Ruckelshaus, G. M. E. Perillo, B. R. Silliman, N. Muthiga, D. Bael and E. Wolanski, *Conserv. Biol.*, 2010, **24**, 207–216.
106. H. Kokko, A. Mackenzie, J. D. Reynolds, J. Lindström and W. J. Sutherland, *Am. Nat.*, 1999, **72**, 358–382.
107. O. Bodin, B. I. Crona and H. Ernstson, *Ecol. Soc.*, 2006, **11**(2).
108. O. Bodin and B. I. Crona, *Global Environ. Change*, 2009, **19**, 366–374.
109. M. Drummond, G. H. Torrance and J. Mason, *Methods for the Economic Evaluation of Health Care Programs,* Oxford University Press, Oxford, 1997.
110. R. Cullen, G. A. Fairburn and K. F. D. Hughey, *Ecol. Econ.*, 2001, **39**, 53–66.
111. K. F. D. Hughey, R. Cullen and E. Moran, *Conserv. Biol.*, 2003, **17**, 93–103.
112. R. Cullen, G. Fairburn and K. Hughey, *Pac. Conserv. Biol.*, 1999, **5**, 115–123.
113. N. Salafsky and R. Margoluis, *Conserv. Biol.*, 1999, **13**, 830–841.
114. E. B. Barbier, *Econ. Policy*, 2007, **22**, 177–229.
115. J. Ehrenfeld, *J. Ind. Ecol.*, 1997, **1**, 41–49.
116. M. T. Brown and S. Ulgiati, *Ecol. Eng.*, 1997, **9**, 51–69.
117. H. T. Odum and E. P. Odum, *Ecosystems*, 2000, **3**, 21–23.
118. S. -L. Huang and W. -L. Hsu, *Landscape Urban Plan.*, 2003, **63**, 61–74.
119. J. A. Foley, R. De Fries, G. P. Asner, C. Barford, G. Bonan, S. R. Carpenter, F. S. Chapin, M. T. Coe, G. C. Daily, H. K. Gibbs, J. H. Helkowski, T. Holloway, E. A. Howard, C. J. Kucharik, C. Monfreda, J. A. Patz, C. Prentice, N. Ramankutty and P. K. Snyder, *Science*, 2005, **309**, 570–574.
120. R. L. Goldman, H. Tallis, P. Kareiva and G. C. Daily, *Proc. Natl. Acad. Sci. U. S. A.*, 2008, **105**, 9445–9448.

Ecosystem Services and Food Production

KEN NORRIS, SIMON G. POTTS AND SIMON R. MORTIMER

ABSTRACT

By 2030, the world's human population could rise to 8 billion people and world food demand may increase by 50%. Although food production outpaced population growth in the 20th century, it is clear that the environmental costs of these increases cannot be sustained into the future. This challenges us to re-think the way we produce food. We argue that viewing food production systems within an ecosystems context provides the basis for 21st century food production. An ecosystems view recognises that food production systems depend on ecosystem services but also have ecosystem impacts. These dependencies and impacts are often poorly understood by many people and frequently overlooked. We provide an overview of the key ecosystem services involved in different food production systems, including crop and livestock production, aquaculture and the harvesting of wild nature. We highlight the important ecosystem impacts of food production systems, including habitat loss and degradation, changes to water and nutrient cycles across a range of scales, and biodiversity loss. These impacts often undermine the very ecosystem services on which food production systems depend, as well as other ecosystem services unrelated to food. We argue that addressing these impacts requires us to re-design food production systems to recognise and manage the limitations on production imposed by the ecosystems within which they are embedded, and increasingly embrace a more multifunctional view of food production systems and associated ecosystems. In this way, we should be able to produce food more sustainably whilst inflicting less damage on other important ecosystem services.

Issues in Environmental Science and Technology, 30
Ecosystem Services
Edited by R.E. Hester and R.M. Harrison
© Royal Society of Chemistry 2010
Published by the Royal Society of Chemistry, www.rsc.org

1 Introduction

By 2030, the world's human population is expected to rise to 8 billion people and world food demand is expected to increase by 50%. In turn, these changes are expected to generate additional needs for water and energy, at a time when climate change is affecting water availability and necessitating a move towards more renewable forms of energy. These issues may come together in what Professor John Beddington, the UK Government's Chief Scientific Advisor, has termed a 'Perfect Storm' of food, water and energy shortages. The complex inter-play of the factors involved are beyond the scope of this review, but these interactions between population growth, food, water and energy mean that food production can no longer be considered in isolation. We need a more holistic framework for producing food that recognises the inter-dependencies with a range of other societal needs.

Many of the issues involved are not new.[1,2] Throughout the 20th century food production outpaced population growth as a result of yield improvements and land clearance for farming. These food production benefits came at a substantial cost to the environment. Agricultural land use has significantly modified global water and nutrient cycles, resulted in the degradation of many of the world's ecosystems and is the major driver of global biodiversity loss.[3–8] Furthermore, there are serious issues about the sustainability of current production systems. For example, in tropical countries soil erosion and the loss of soil fertility are major constraints on food production.[9–11] In developed countries, most production systems rely on artificial surrogates to manage key processes, such as pollination, pest and disease control, and there is concern that current threats to these processes (*e.g.* dramatic declines in domestic bees) could negatively affect crop yields.[12] The scale of these impacts is perhaps not surprising, given that cultivated systems cover about 25% of the Earth's land surface.[13]

All of these issues reflect a failure to recognise that food production systems are embedded within an ecosystem. This ecosystem might be highly managed, such as the agro-ecosystems typical of many intensively managed agricultural landscapes in Europe, North America and Australasia, or the ecosystem might retain a considerable amount of its original structure and function, such as many small-holder farming systems in developing countries or the harvesting of wild species. Nevertheless, recognition that food production should be viewed within an ecosystem context leads to two important insights (see Figure 1). First, food production is potentially dependent on a wide range of services provided by the ecosystem. These include plant and animal diversity from which crops and livestock are derived, soil and water resources, nutrient cycling, biomass production, pollination and pest control, and so on. Second, food production systems have implications for the ecosystem of which they are part through, for example, habitat loss and degradation, water use, diffuse pollution, biodiversity loss, and so on. Understanding these feedbacks between the food production system and the ecosystem within which it is embedded is the key to developing a more holistic framework for producing food that also

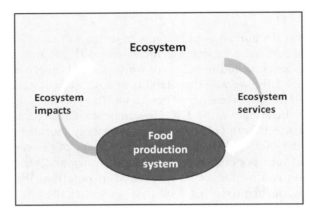

Figure 1   Viewing food production systems within an ecosystems context.

recognises other major challenges we face, such as population growth, water and energy use.

In this paper we consider food production within an ecosystems context. We have two main aims:

1. To provide an overview of the ecosystem services that underpin food production systems, without which yields may be substantially reduced.
2. To highlight the important ecosystem impacts associated with food production systems.

In this way, we will explore the feedback between food production systems and the wider ecosystems within which these are embedded. Before addressing these issues in detail, we outline the type of framework typically used to describe a range of ecosystem services. Our paper concludes with some perspectives on the future.

2   Ecosystem Services Important for Food Production

2.1   A Conceptual Framework

The Millennium Ecosystem Assessment (MEA), published in 2005, provided for the first time an overview of the status of the World's ecosystems, the delivery of ecosystem services and the potential consequences for human well-being now and into the future.[13] It recognised that everyone depends on the services provided by the World's ecosystems, such as food, water, disease management, climate regulation, spiritual fulfilment and aesthetic enjoyment. It concluded that although ecosystem management over the last 50 years had benefited human well-being through increased outputs of food, fresh water, timber, fibre and fuel, the full costs associated with these gains are only now becoming apparent.

The MEA recognised four broad categories of ecosystem services – provisioning, regulating, cultural and supporting. *Provisioning services* are the products obtained from ecosystems, which includes food, fibre and fuel. *Regulating services* are the benefits arising from the regulation of ecosystem processes, such as climate regulation, water purification, pollination and the control of pests and diseases. *Cultural services* are the non-material benefits people obtain from ecosystems, such as spiritual enrichment, recreation and aesthetic experiences. *Supporting services* are those services necessary for the production of all other ecosystem services, such as soil formation and nutrient cycling. A number of conceptual frameworks for ecosystem services have been developed since the MEA,[14,15] motivated by a need to avoid double counting when attempting to link services to the values and benefits people obtain from them.[16] Nevertheless, the MEA categories do provide a reasonable basis for recognising the inter-dependencies between ecosystem services that are essential to food production, so we have used MEA terminology in the following discussion.

Food is a provisioning service that includes crop and livestock production, aquaculture and the harvesting of wild nature. These services are themselves dependent on a range of provisioning, regulating and supporting services. All food production systems based on agriculture depend on the genetic diversity (a provisioning service) of crop plants and animals to directly provide food products, as well as soil and water resources (a mixture of regulating and supporting services) to support the associated production systems. Consequently, we recognise this suite of services as general ecosystem services to agriculture and discuss them as a group. Each food production service – crops, livestock, aquaculture and harvesting wild nature – has its own dependencies on other services, so we highlight the key relationships involved for each food production service.

2.2 Ecosystem Services

2.2.1 General Services to Agriculture
In this section, we consider three major types of resources required for agriculture, namely the genetic resources necessary for the development of food crops and livestock breeds, fresh water and fertile soil to support primary production. Agricultural production is dependent on a range of ecosystem services which relate to these three resources. For each of these three areas we discuss the nature and importance of the service, its relevance to agriculture and current trends in the level of service provision.

Genetic Resources. Agriculture by its very nature involves a simplification of biodiversity through the selective cultivation or rearing of a limited number of species. The bulk of world food production comes from a limited range of plant and animal species. The thirty most widely grown crops provide about 90% of global calorific value, with three of these (wheat, rice and maize) contributing an estimated 50% of intake, whilst fewer than 14 species provide an estimated 90% of global livestock production.[13] There is considered to be a large potential for the development and improvement of underutilised species[17] which may contribute not only to sustainable agricultural production in the face of changing environmental conditions, but also to enhancing human health through a diversification of diets.

Genetic diversity contributes to agriculture not only through providing the raw material for breeding, but also through conferring protection against pests and diseases. In extensive systems using so-called land races, within-field diversity in genetic composition of the crop species provides an insurance mechanism against pest and diseases. The development of modern agricultural varieties has led to a loss of such within-field diversity, although agricultural practices such as rotation or inter-annual variation in varieties grown offer similar protection. Loss of biodiversity, either of crop relatives of other species, poses a threat by limiting the pool of traits available for breeding programmes. Thus there is an insurance value in biodiversity conservation in providing a bank of genes or traits that may serve future agricultural breeding programmes.[18]

Water Resources. Agriculture is dependent on the provision of fresh water of sufficient quantity and quality, and at appropriate times of the year to support crop and livestock production. The majority of the world's crop systems are rain fed, although crops from irrigated systems comprise 18% of cropped area and 40% of economic value of crop production worldwide. It is estimated that approximately 70% of global freshwater consumption is for agriculture.[13]

Clearly predictability in both the timing and stability of supply confer an advantage to farmers. Changes in climate, in places exacerbated by regional changes in land use, affect the amount of precipitation supplying rain-fed agricultural systems, but it is amongst irrigated systems that water supply issues are most sensitive to environmental change. Within catchments, land use change upstream of agricultural areas affects the quantity, quality and timing of water resources supplied. Forests stabilise supply and quality, with deforestation leading to soil erosion, greater sediment loads and increased variability in flow rates downstream.

Total water demand for agriculture is increasing globally. In many countries emphasis on self sufficiency of food supply has exacerbated problems of water scarcity. In many areas the demand for water for agriculture is predicted to outstrip supply, in spite of increases in water use efficiency. In the Middle East, North Africa, sub-Saharan Africa and parts of Asia extraction of water for agriculture represents 85–90% of water withdrawals. Irretrievable losses of water from irrigated systems are estimated to account for one third of total freshwater use.[13]

Clearly there are conflicts between agriculture and other users of water downstream, whether for food production, drinking water or wetland protection. Promotion of sustainable land management by farmers and integrated water management at the catchment scale offer solutions to the conflicts between agriculture and other sectors for water resources. For agriculture, improving water use efficiency through breeding, agricultural management practices (*e.g.* mulching, contour farming, reducing losses to soil evaporation, weeds or run off), water harvesting, water recycling and use of marginal quality water offer solutions.[19]

Soil Resources. Soils supply the elements necessary for primary production and consequently support both crop and livestock systems. Soil structure and fertility determine the suitability of land for agriculture and the quantity and quality of agricultural outputs. Soil biota performs essential ecosystem services of soil formation and nutrient cycling, the movement of elements between various abiotic and biotic forms. Soil biota regulates the rates of flow between different forms, the sizes of pools and consequently the availability of nutrients to support plant growth.

Soil organic matter is a key component of soil, influencing a number of ecosystem services including nutrient cycling, water retention, soil structure and erosion.[20] Agriculture utilises the store of energy and nutrients contained within soil organic matter to support crop growth, resulting in a lower equilibrium between organic matter supply and production. There is therefore a trade off between food production and the other ecosystem services influenced by the level of organic matter in the soil. Recent trends towards intensification of agriculture, specialisation by farmers on particular products and urbanisation of the population have resulted in more open cycles of nutrients, with reductions in the return of crop residues, animal dung and other waste products to the soils supporting food production.

In agriculture, nutrient cycling is often supplemented by agricultural practice, most notably the use of industrial fertilizers. Nutrient cycles have been substantially altered by human activity since start of industrial period. Annual fluxes of nitrogen are estimated to be more than double the rates of 200 years ago. Only a fraction of the amount fixed is denitrified back to gaseous atmospheric nitrogen. Phosphorus cycles have shown a similar increase in annual flux, largely as the result of use of mined phosphorus for agriculture. Surplus nitrogen and phosphorus accumulates in land and water leading to eutrophication problems. However, whilst nutrient accumulation is a problem in some industrialised areas of the world, in other areas unsustainable agricultural practices have led to impoverishment of soil fertility.

In modern intensive agriculture it is typical to replace ecosystem services provided by biodiversity with equivalents derived from human labour and or petrochemical energy or its products.[21] This has the effect of replacing one form of risk (relying on naturally provided ecosystem services) with another (reliance of the markets for labour and technology).[22] A number of agricultural land management practices can contribute to sustainable soil management and help to promote delivery of ecosystem services including food production.[23] These include use of organic fertilizers and soil amendments; modifications to tillage practice; incorporation of leguminous crops or cover crops into rotations; and landscape level management practices to minimise soil erosion (*e.g.* appropriate directional placement of row crops, use of hedgerows and banks).

2.2.2 Crop Production
Crop production is a provisioning service, which in turn is highly dependent upon two regulating services: pollination and pest regulation.[13] Both regulating services can be provided biotically, in which case they are characterised as being

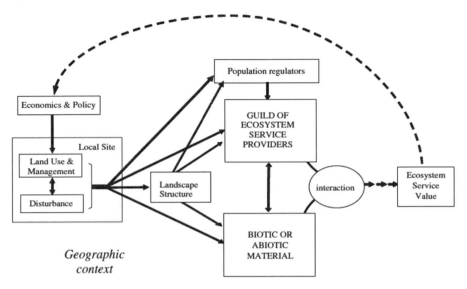

Figure 2   A generalized conceptual framework for mobile agent-based ecosystems services, such as pollination and biocontrol.[24] Land-use change impacts on the interaction of ecosystem service providers (*e.g.* pollinators and natural enemies) with the biotic (*e.g.* flowers and pests) and abiotic environment to deliver services (*e.g.* pollination and biocontrol). The value of the services can then feedback through policy and market forces to modify land-use and management.

mobile-agent-based ecosystem services where the service providing units are usually insects (see Figure 2).[24]

Pollination is the process of pollen transfer from male parts of a flower (anthers) to female parts (stigma) and is essential for crops where the harvested product is a fruit, seed or nut. Pollen transfer can be 'abiotic', where the vector is wind, water or gravity (typical of most staple crops, such as cereals and rice), or 'biotic', where the vector is an animal such as an insect, bird or mammal (characteristic of the majority of fruit crops). Approximately 75% of all global crops used for human consumption are reliant upon insect pollination, mostly by bees,[25] and the annual value of biotic pollination services is estimated to be €153 billion, which represents almost 10% of the total economic value of world agriculture.[26] The fraction of global agriculture which depends on biotic pollination has increased by $>300\%$ since 1961, which has outstripped the increase ($\sim 45\%$) in managed honey bee colonies,[27] leading to concerns about future shortfalls in services. While worldwide, there has been an increase in managed honey bees, severe regional declines have been documented in Europe[28] and the USA,[29] where the parasitic mite *Varroa destructor* has virtually wiped out all feral honey bee colonies.[30,31]

While managed honey bees pollinate many entomophilous crops, unmanaged bees and other insects ('wild' pollinators) account for an important, but as yet unquantified proportion of biotic crop pollination.[25] Furthermore several crops rely on wild bees as they cannot be efficiently pollinated by honeybees owing to morphological and/or behavioural restrictions.[32] The diversity of bee pollinators has been found to be positively correlated with the delivery of pollination services in some crops.[12,33] However, wild pollinators are also reported to be in serious decline in several countries[34,35] and though data is lacking for many parts of the world, the phenomena is expected to be widespread given that many of the known drivers of pollinator loss are also widespread. For instance, the loss or fragmentation of semi-natural habitats around farms and agricultural intensification are known to cause local losses of diversity and abundance in pollinator communities,[36] and this may also translate into loss of pollination services in some cases.[12,37] Other drivers of pollinator loss include pesticides,[38] pathogens,[39] climate change[40] and the introduction of alien species.[41]

Biological control is the management of natural enemies for the regulation of pest densities in agricultural systems. In addition to decreasing crop damage by pests, biological control can reduce the insecticide resistance of pests and minimise the need for chemical control methods, such as pesticide sprays.[42] Natural enemies can be predators (*e.g.* ladybirds), parasites (*e.g.* some fly larvae), parasitoids (*e.g.* parasitic wasps) or pathogens (*e.g.* some fungi, bacteria and viruses); all these taxa can be potentially used to control populations of agricultural pests including animals, weeds and diseases.[43] Worldwide it is estimated that 95% of approximately 100 000 species of potential arthropod pests in agricultural fields and forests are regulated by natural enemies.[42] The value of biological control services to global crop production was estimated to be worth US$400 billion per annum;[44] more recent estimates value the annual contribution of arthropods natural enemies to crop production in the USA at US$4.5 billion.[45]

Biological control services can be achieved through three main approaches:

1. *Conservation Control* relies on native natural enemies to control native pests. The local availability and effectiveness of natural enemies can be improved by providing food resources (such as nectar flowers for aphidophagous hoverflies and parasitic wasps);[46] supplying alternative prey and hosts;[47] creating overwintering refuges (*e.g.* beetle banks);[48] and modifying the microhabitats of crops to be more suitable for natural enemies.[49]

2. *Classical Control*, sometimes known as innoculative control, involves the introduction of an exotic natural enemy to control an exotic pest which has become established on a crop where it was not previously found. It is thought to be practiced in ~ 10% of cultivated land globally.[42] Classical control is often most effective in perennial cropping systems where the longer-term benefits can be achieved through the persistence of natural enemy populations, for example, the introduction of a wasp parasitoid (*Aphelinus mali*) to control of the introduced woolly apple aphid (*Eriosoma lanigerum*).[50]

3. *Augmentation Control* relies on the periodic release of supplemental natural enemies to control pests. Unlike classical control, highly disturbed annual cropping systems may not allow natural enemy populations to persist between cropping cycles, and so the control agents generally need to be commercially mass produced and deployed at regular intervals. Augmentative control is applied to <0.05% of the worlds cultivated land.[51] An example of this approach is the control of the cotton bollworm (*Heliothis virescens*) by *Trichogramma* wasps.[51] Biological control is a key component of integrated pest management, which is a toolkit of complementary strategies including biological, physical and chemical methods which aim to regulate pest populations to keep them at an acceptable level while significantly reducing pesticide use.

2.2.3 Livestock Production

Livestock production takes place in managed and natural grasslands and so depends ultimately on the primary productivity of grassland ecosystems (a supporting ecosystem service) and associated ecological processes. Humans harvest about 30% of global net primary production (NPP) annually, of which about one third is directly grazed by livestock.[52] Global livestock production is expected to double by 2050,[53] placing additional demands on NPP. Many managed grassland systems in developed temperate regions are often highly simplified, being based on a small number of grass species, and intensively managed through the use of fertilizers. Dependence on ecosystem services in these systems is limited, with food production being almost entirely dependent on artificial surrogates that have replaced the ecological processes that underpin primary productivity in natural grassland ecosystems. Livestock production also depends on animal feeds derived from crops. Livestock production relies, therefore, indirectly on a wide range of ecosystem services that underpin crop production (see sections 2.2.1 and 2.2.2).

2.2.4 Inland Fisheries and Aquaculture

These are important food production systems, accounting for about a quarter of global fisheries production. There are three main types of system: subsistence fisheries that are particularly important in parts of Africa and SE Asia; commercial fisheries based predominantly on the lakes of North America and Africa; and recreational fisheries that have high economic value but play a very limited role in food production.

In a broad sense, subsistence fisheries are supported by ecosystem services in a very similar way to marine fisheries (see section 2.2.5). They depend on aquatic ecosystems and associated processes, and are vulnerable to perturbations (*e.g.* changes in water management, pollution, land-use change and pollution) that disrupt the processes that under-pin the biomass of fish available to be harvested. In contrast, commercial fisheries are becoming increasingly reliant on a few species (some of which are non-native) and are based on more intensive management practices (such as feeding and restocking). In this sense, they are becoming increasingly like crop and livestock systems on land, and as such depend on a range of comparable supporting (*e.g.* primary productivity, nutrient cycling) and regulating (*e.g.* water purification, disease control) services. There is also a growing trend, as on land, to replace natural ecosystem services with artificial surrogates (*e.g.* feeding and restocking).

2.2.5 Marine Fisheries

Marine fisheries are a globally important source of food – they are a significant source of protein for nearly half the World's people. Marine fisheries also affect food production indirectly through the provision of fish and animal feeds to aquaculture and livestock production systems. Demand for fish is increasing.

Unlike crop and livestock systems that often depend on services from highly managed ecosystems, marine fisheries are the product of largely natural coastal and marine ecosystems. As such, they depend on food webs, and the flow of energy and nutrients through these, which ultimately determines the biomass of fish available for harvest. In turn, food webs and associated flows depend on the integrity of key habitats and populations of key species. There is a wealth of evidence showing that the perturbation of these ecosystem processes through habitat loss, pollution and climate change can adversely affect fish populations,[54] as well as ecosystem feedbacks caused by the removal of the fish themselves.[55] Finally, there is evidence that species diversity positively affects fishery harvests and resilience to exploitation,[56] implying that fisheries depend to some extent on a range of evolutionary and ecological processes that determine wild species diversity.

2.2.6 Terrestrial Wild Animal Products

A wide range of terrestrial animals are harvested for food, including mammals, birds, reptiles, amphibians and invertebrates (*e.g.* snails, insects and insect products such as honey), but large-bodied mammals, particularly ungulates, make up the majority of the biomass harvested. Locally, particularly in forested regions of the tropics, wild animals are a significant source of protein.[57] The overall perception is that these sources of food are decreasing in importance because of the ongoing loss of forest ecosystems that support many of the harvested species, but data on harvesting are sparse, harvesting behaviour varies (*e.g.* consuming locusts during plagues) and trade in some wild animal products is increasing (*e.g.* bushmeat). As a result, the overall picture is unclear.

As in fisheries based on harvesting wild species, food derived from terrestrial wild animals depends on ecosystems, typically forest, and the associated processes that under-pin the biomass available to be harvested. As a result, biomass varies spatially in relation to changes in key ecosystem services, such as local water and nutrient cycling (supporting services).[58,59] Food derived from wild animals is susceptible to perturbations (such as forest clearance) that affect ecosystem integrity and the associated processes.

3   The Impact of Food Production on Ecosystems

It is quite clear from the above discussion that a wide range of managed and natural food production systems are critically dependent on ecosystem services, particularly the supporting and regulating services. In turn, food production systems have ecosystem impacts, with implications for ecosystem services that go beyond simply those services linked to the food production system itself. These impacts occur through a range of processes including habitat loss and degradation, changes in water and nutrient cycles, and biodiversity loss. In this section, we discuss these impacts and their consequences.

Cultivated systems cover 25% of the Earth's terrestrial land surface.[13] The footprint of food production systems on ecosystems is, however, much more extensive than this as it covers any terrestrial, coastal or marine ecosystem in which food production occurs through farming or harvesting activities, plus ecosystems linked to these by ecosystem processes. Wherever there are people, local ecosystems are modified by food production activities. Farmland is derived by clearing natural ecosystems, so the global impact of crop and livestock production systems is huge, and it is the major driver of ecosystem change in many regions. In some developed countries, such as the UK, agriculture together with other land management activities has replaced almost all the original, native ecosystems. Comparable impacts are evident in marine ecosystems.[55,60] Estimates suggest that around 20% more land will be required in developing countries for food production by 2050,[61] so that these ecosystem impacts are set to continue.

Food production systems have modified water and nutrient cycles at multiple scales.[1,2] These modifications often have implications beyond the ecosystems in which they originally occurred, through large-scale changes to regional water and nutrient cycles. For example, removing forest ecosystems can alter the regional water cycle and associated rainfall in areas beyond the forest zone;[62] waterborne nutrients from agriculture have caused phytoplankton blooms in marine ecosystems;[63] and agriculture plays a major role in the global carbon cycle and hence in climate change.[64]

Food production systems are also a major driver of biodiversity loss, particularly in terrestrial ecosystems.[7,65] This mainly occurs because food production systems typically only retain a fraction of the biodiversity found in the natural ecosystem from which they were derived. This fraction varies between food production systems, being highest in systems that retain characteristics of the original ecosystem (such as silvopastoral and agro-forestry production systems), but some biodiversity loss is typical.[66] Biodiversity loss has also been documented in recent years in highly managed food production systems found in Europe and elsewhere as a result of management changes designed to boost production.[67,68] Food production systems also modify interactions between species,[69] potentially modifying ecosystem processes and the services of which these are part.[70]

The impact of food production systems on ecosystems is global and profound. What are the consequences of these impacts for ecosystem services? We address this question from two perspectives. First, we consider the implications of these impacts on ecosystem services that under-pin food production. Next, we consider ecosystems more broadly and the developing agenda that considers ecosystems as *multi-functional* systems, rather than focusing solely on traditional, sectoral interests (such as farming) that relate to a specific ecosystem service.

The ecosystem degradation caused by food production systems often undermines the provision of the ecosystem services required to sustain them.[20] There are numerous examples. Over-grazing of grassland ecosystems by livestock causes ecosystem degradation that reduces primary production (a supporting service) and hence the ability of the ecosystem to sustain future livestock production. Soil erosion and the loss of soil fertility frequently occur because cropping systems expose the soil to weathering and deplete nutrients. This degrades key supporting services, such as nutrient cycling, which in turn reduces the ability of the ecosystem to support future crop production. The removal of natural habitats to increase the area available for crop production often causes biodiversity loss, which can degrade key regulating services (such as pollination and pest control) dependent on biodiversity. As a result, food production systems are dependent on artificial surrogates, such as honey bees and chemical pesticides, without which yields may decline. Finally, there are a number of examples of over-exploitation in which the biomass of animals being harvested from ecosystems exceeds the ability of the ecosystem to support this biomass removal.[55,71]

In all of these examples, food production is threatened because the food production system itself is over-exploiting the ecosystem services on which it depends. In this ecosystems context, the sustainability of the food production system depends on the extent to which it operates within the limits imposed by the ecosystem within which it is embedded. Of course, natural ecosystem services can and often are supplemented or replaced entirely by artificial surrogates (*e.g.* fertilizers, pesticides, domesticated pollinators), but this strategy contains risks if the artificial system fails and there is no natural back-up system to replace it. These issues are currently very real with respect to pollination services in Europe and North America (see section 2.2.2). In fact, recent work in the US has shown that restoring natural pollination services can have economic benefits in terms of increased crop production.[72] In the long-term, it is essential to ensure that we continue to have the ecosystem services available that our food production systems require. It is standard practice in engineered systems, from aeroplanes to buildings, to design systems that minimize the risk of failure. We need the same approach to food production systems and their associated ecosystem services.

If food production is considered as an ecosystem service, then food production systems represent ecosystem management tools designed for the delivery of one specific provisioning service. The important question then becomes if we manage or exploit ecosystems for food production, to what extent do we compromise the ability of those ecosystems to deliver a range of other ecosystem services? The examples discussed above suggest that in general terms detrimental impacts on other ecosystem services are common. For example, if a large area of tropical forest is cleared to provide land for a crop production system (such as soya, maize or sugar cane), a number of ecosystem services are likely to be adversely affected (including water and nutrient cycling, climate regulation, water purification, erosion regulation, and the diversity of wild species). Nevertheless, our quantitative understanding of such trade-offs in particular ecosystems remains relatively poor in terms of both pattern and process. Improving our understanding is very much at the cutting edge of current ecosystem science.

Although much remains to be done in terms of the science, the growing recognition that we have to define, quantify and manage trade-offs between services in ecosystems is becoming embodied within a *multi-functional* ecosystems approach.[66,73] This is a holistic approach in which the full range of consequences of a particular ecosystem system change are explored, valued and managed. In principle, this approach should allow us to identify any adverse changes and adapt ecosystem management measures to avoid them. This can only be done, however, if we understand not only the services being provided by ecosystems, but also the values and benefits people derive from these services.[14,16] This remains a substantial challenge, not least because it requires us to value services for which there are no markets, as well as services (such as food production) for which markets exist. There is also the challenge of providing the ecological information necessary to support emerging markets for ecosystem services (*e.g.* carbon credits), understanding how these might work, and how they might affect the way ecosystems are managed and the communities involved. Meeting these challenges will require natural and social scientists to increasingly work together, and develop unified concepts within an overall ecosystems framework.[15,74] Over the coming years, this ecosystems approach could completely re-shape the way we consider food production systems.

4   Conclusions

The Millennium Ecosystem Assessment concluded that although ecosystem management over the last 50 years had benefited human well-being through increased outputs of food, fresh water, timber, fibre and fuel, the full costs associated with these gains are only now becoming apparent. In the context of food production, we have designed increasingly efficient food production systems that have enabled food production to keep pace with increasing and changing demands from a growing human population. Technological developments in animal and plant breeding, agro-chemicals, mechanisation and integrated farming systems have played a huge role in these productivity gains. Nevertheless, we have shown that most, if not all, food production systems rely on a wide range of ecosystem services that have been increasingly exploited during these productivity gains. While the role of technological development is widely recognised, the important role played by ecosystem services is poorly understood by many people and frequently overlooked.

In an ecosystems context, what the MEA calls the 'full costs associated with these gains' represent degradation in the ecosystem within which a food production system is embedded and in other ecosystems linked to it by ecosystem processes. These 'costs' have two main consequences. First, we have shown that ecosystem degradation caused by food production systems can under-mine the very ecosystem services on which food production systems depend. This either reduces productivity, or means that continued productivity is dependent on artificial surrogates. We argue that these outcomes are undesirable and risky. Second, food production systems can damage a range of ecosystem services related to societal needs other than food, and we give examples. These impacts can be transferred to ecosystems and people outside the food production system causing the damage. In our view, these costs mean that current food production systems should be regarded as predominantly unsustainable.

Addressing these issues requires us to embrace an ecosystems approach, and apply it in two main ways. First, we need to develop food production systems that recognise and manage the limits imposed by the ecosystems in which they are embedded. This means improving our understanding of ecosystem processes and associated services with which food production systems interact through management practices. This will require a new, integrated research agenda between the agricultural and environmental sciences. It will also require us to re-think the way we develop and use technology, away from the current position in which it is used to largely replace key ecosystem services and towards one in which technology plays a role as a tool to manage ecosystem services. Second, we need to develop our understanding of trade-offs between food production and other key ecosystem services, and design novel policy and management measures to deal with important trade-offs. Again, this will require new interdisciplinary research. This *multi-functional* view also has challenging implications for policy and practice that still tends to be rather isolated within the traditional areas of agriculture, fisheries, food and environment.

Whether or not you feel a 'Perfect Storm' is likely, it serves as an important metaphor highlighting the need for change. An ecosystems approach provides us with the integrated, holistic framework we need to manage our planet for a range of societal needs. Feeding a growing human population is obviously critically important. In our view, this can only be done by recognising and embracing the concept that food production systems are embedded within ecosystems. They depend on ecosystem services and have ecosystem impacts. This view does not under-value the role of technology, but would argue that its role should now be considered within an ecosystems context. We argue that the benefits of this ecosystem view will be more sustainable food production systems and less damage to other important ecosystem services. To borrow a phrase, 'It's the ecosystem, stupid!'

References

1. J. A. Foley, G. P. Asner, C. Barford, G. Bonan, S. R. Carpenter, F. S. Chapin, M. T. Coe, G. C. Daily, H. K. Gibbs, J. H. Helkowski,

T. Holloway, E. A. Howard, C. J. Kucharik, C. Monfreda, J. A. Patz, I. C. Prentice, N. Ramankutty and P. K. Snyder, *Science*, 2005, **309**, 570–574.

2. D. Tilman, *Proc. Natl. Acad. Sci. U. S. A.*, 1999, **96**, 5995–6000.
3. A. Bondeau, P. C. Smith, S. Zaehle, S. Schaphoff, W. Lucht, W. Cramer and D. Gerten, *Global Change Biol.*, 2007, **13**, 679–706.
4. W. Junkermann, J. Hacker, T. Lyons and U. Nair, *Atmos. Chem. Phys.*, 2009, **9**, 6531–6539.
5. J. N. Galloway, A. R. Townsend, J. W. Erisman, M. Bekunda, Z. C. Cai, J. R. Freney, L. A. Martinelli, S. P. Seitzinger and M. A. Sutton, *Science*, 2008, **320**, 889–892.
6. J. G. Van Minnen, K. K. Goldewijk, E. Stehfest, B. Eickhout, G. van Drecht and R. Leemans, *Climat. Change*, 2009, **97**, 123–144.
7. R. E. Green, S. J. Cornell, J. P. W. Scharlemann and A. Balmford, *Science*, 2005, **307**, 550–555.
8. R. Fujimaki, A. Sakai and N. Kaneko, *Ecol. Res.*, 2009, **24**, 955–964.
9. R. Lal, *Science*, 2004, **304**, 1623–1627.
10. S. Ngoze, S. Riha, J. Lehmann, L. Verchot, J. Kinyangi, D. Mbugua and A. Pell, *Global Change Biol.*, 2008, **14**, 2810–2822.
11. M. A. Stocking, *Science*, 2003, **302**, 1356–1359.
12. C. Kremen, N. M. Williams and R. W. Thorp, *Proc. Natl. Acad. Sci. U. S. A.*, 2002, **99**, 16812–16816.
13. MEA, *Ecosystems and Human Well-Being: Synthesis*, World Resources Institute, Island Press, Washington D.C., 2005.
14. B. Fisher, K. Turner, M. Zylstra, R. Brouwer, R. de Groot, S. Farber, P. Ferraro, R. Green, D. Hadley, J. Harlow, P. Jefferiss, C. Kirkby, P. Morling, S. Mowatt, R. Naidoo, J. Paavola, B. Strassburg, D. Yu and A. Balmford, *Ecol. Appl.*, 2008, **18**, 2050–2067.
15. S. R. Carpenter, H. A. Mooney, J. Agard, D. Capistrano, R. S. DeFries, S. Diaz, T. Dietz, A. K. Duraiappah, A. Oteng-Yeboah, H. M. Pereira, C. Perrings, W. V. Reid, J. Sarukhan, R. J. Scholes and A. Whyte, *Proc. Natl. Acad. Sci. U. S. A.*, 2009, **106**, 1305–1312.
16. B. Fisher and R. K. Turner, *Biol. Conserv.*, 2008, **141**, 1167–1169.
17. R. L. Naylor, W. P. Falcon, R. M. Goodman, M. M. Jahn, T. Sengooba, H. Tefera and R. J. Nelson, *6th International Conference of the International Consortium on Agricultural Biotechnology Research*, Ravello, Italy, 2002.
18. R. Hajjar, D. I. Jarvis and B. Gemmill-Herren, *Agri. Ecosyst. Environ.*, 2008, **123**, 261–270.
19. R. J. Thomas, *J. Environ. Monit.*, 2008, **10**, 595–603.
20. W. Zhang, T. H. Ricketts, C. Kremen, K. Carney and S. M. Swinton, *Ecol. Econ.*, 2007, **64**, 253–260.
21. M. G. Kibblewhite, K. Ritz and M. J. Swift, *Philos. Trans. R. Soc. London, Ser. B*, 2008, **363**, 685–701.
22. M. J. Swift, A. M. N. Izac and M. van Noordwijk, *Agri. Ecosyst. Environ.*, 2004, **104**, 113–134.

23. K. E. Giller, M. H. Beare, P. Lavelle, A. M. N. Izac and M. J. Swift, *Workshop on Agricultural Intensification, Soil Biodiversity and Agroecosystem Function in the Tropics*, Hyderabad, India, 1995.

24. C. Kremen, N. M. Williams, M. A. Aizen, B. Gemmill-Herren, G. LeBuhn, R. Minckley, L. Packer, S. G. Potts, T. Roulston, I. Steffan-Dewenter, D. P. Vazquez, R. Winfree, L. Adams, E. E. Crone, S. S. Greenleaf, T. H. Keitt, A. M. Klein, J. Regetz and T. H. Ricketts, *Ecol. Lett.*, 2007, **10**, 299–314.

25. A. M. Klein, B. E. Vaissiere, J. H. Cane, I. Steffan-Dewenter, S. A. Cunningham, C. Kremen and T. Tscharntke, *Proc. R. Soc. London, Ser. B*, 2007, **274**, 303–313.

26. N. Gallai, J. M. Salles, J. Settele and B. E. Vaissiere, *Ecol. Econ.*, 2009, **68**, 810–821.

27. M. A. Aizen and L. D. Harder, *Curr. Biol.*, 2009, **19**, 915–918.

28. S. G. Potts, S. P. M. Roberts, R. Dean, G. Marris, M. Brown, Jones R. and J. Settele, *J. Apic. Res.*, in press.

29. NRC, *Status of Pollinators in North America*, National Research Council, 2006.

30. B. Kraus and R. E. Page Jr, *Environ. Entomol.*, 1995, **24**, 1473–1480.

31. R. F. A. Moritz, F. B. Kraus, P. Kryger and R. M. Crewe, *J. Insect Conserv.*, 2007, **11**, 391–397.

32. P. G. Willmer, A. A. M. Bataw and J. P. Hughes, *Ecol. Entomol.*, 1994, **19**, 271–284.

33. P. Hoehn, T. Tscharntke, J. M. Tylianakis and I. Steffan-Dewenter, *Proc. R. Soc. London, Ser. B*, 2008, **275**, 2283–2291.

34. J. C. Biesmeijer, S. P. M. Roberts, M. Reemer, R. Ohlemuller, M. Edwards, T. Peeters, A. P. Schaffers, S. G. Potts, R. Kleukers, C. D. Thomas, J. Settele and W. E. Kunin, *Science*, 2006, **313**, 351–354.

35. P. H. Williams and J. L. Osborne, *Apidologie*, 2009, **40**, 367–387.

36. R. Winfree, R. Aguilar, D. P. Vazquez, G. LeBuhn and M. A. Aizen, *Ecology*, 2009, **90**, 2068–2076.

37. T. H. Ricketts, J. Regetz, I. Steffan-Dewenter, S. A. Cunningham, C. Kremen, A. Bogdanski, B. Gemmill-Herren, S. S. Greenleaf, A. M. Klein, M. M. Mayfield, L. A. Morandin, A. Ochieng, S. G. Potts and B. F. Viana, *Ecol. Lett.*, 2008, **11**, 1121–1121.

38. C. A. Brittain, M. Vighi, R. Bommarco, J. Settele and S. G. Potts, *Basic Appl. Ecol.*, in press.

39. D. L. Cox-Foster, S. Conlan, E. C. Holmes, G. Palacios, J. D. Evans, N. A. Moran, P. L. Quan, T. Briese, M. Hornig, D. M. Geiser, V. Martinson, D. vanEngelsdorp, A. L. Kalkstein, A. Drysdale, J. Hui, J. H. Zhai, L. W. Cui, S. K. Hutchison, J. F. Simons, M. Egholm, J. S. Pettis and W. I. Lipkin, *Science*, 2007, **318**, 283–287.

40. C. F. Dormann, O. Schweiger, P. Arens, I. Augenstein, S. Aviron, D. Bailey, J. Baudry, R. Billeter, R. Bugter, R. Bukacek, F. Burel, M. Cerny, R. De Cock, G. De Blust, R. DeFilippi, T. Diekotter, J. Dirksen, W. Durka, P. J. Edwards, M. Frenzel, R. Hamersky, F. Hendrickx, F. Herzog,

S. Klotz, B. Koolstra, A. Lausch, D. Le Coeur, J. Liira, J. P. Maelfait, P. Opdam, M. Roubalova, A. Schermann-Legionnet, N. Schermann, T. Schmidt, M. J. M. Smulders, M. Speelmans, P. Simova, J. Verboom, W. van Wingerden and M. Zobel, *Ecol. Lett.*, 2008, **11**, 235–244.
41. D. Thomson, *Ecology*, 2004, **85**, 458–470.
42. J. S. Bale, J. C. van Lenteren and F. Bigler, *Philos. Trans. R. Soc. London, Ser. B*, 2008, **363**, 761–776.
43. T. S. Bellow and T. W. Fisher, *Handbook of Biological Control*, Academic Press, San Diego CA., 1999.
44. R. Costanza, R. d'Arge, R. de Groot, S. Farber, M. Grasso, B. Hannon, K. Limburg, S. Naeem, R. V. O'Neill, J. Paruelo, R. G. Raskin, P. Sutton and M. van den Belt, *Nature*, 1997, **387**, 253–260.
45. D. A. Landis, M. M. Gardiner, W. van der Werf and S. M. Swinton, *Proc. Natl. Acad. Sci. U. S. A.*, 2008, **105**, 20552–20557.
46. F. J. J. A. Bianchi and F. L. Wackers, *Biol. Control*, 2008, **46**, 400–408.
47. E. W. Evans, *Eur. J. Entomol.*, 2008, **105**, 369–380.
48. K. L. Collins, N. D. Boatman, A. Wilcox and J. M. Holland, *Agric. Ecosyst. Environ.*, 2003, **96**, 59–67.
49. J. D. Harwood, K. D. Sunderland and W. O. C. Symondson, *J. Animal Ecol.*, 2003, **72**, 745–756.
50. S. K. Asante, *Plant Protect. Quarterly*, 1997, **12**, 166–172.
51. J. C. Van Lenteren and V. H. P. Bueno, *Biocontrol*, 2003, **48**, 123–139.
52. H. Haberl, K. H. Erb, F. Krausmann, V. Gaube, A. Bondeau, C. Plutzar, S. Gingrich, W. Lucht and M. Fischer-Kowalski, *Proc. Natl. Acad. Sci. U. S. A.*, 2007, **104**, 12942–12945.
53. R. C. Ilea, *J. Agric. Environ. Ethics*, 2009, **22**, 153–167.
54. D. R. Bellwood, T. P. Hughes, C. Folke and M. Nystrom, *Nature*, 2004, **429**, 827–833.
55. D. Pauly, V. Christensen, J. Dalsgaard, R. Froese and F. Torres, *Science*, 1998, **279**, 860–863.
56. B. Worm, E. B. Barbier, N. Beaumont, J. E. Duffy, C. Folke, B. S. Halpern, J. B. C. Jackson, H. K. Lotze, F. Micheli, S. R. Palumbi, E. Sala, K. A. Selkoe, J. J. Stachowicz and R. Watson, *Science*, 2006, **314**, 787–790.
57. J. E. Fa, D. Currie and J. Meeuwig, *Environ. Conserv.*, 2003, **30**, 71–78.
58. C. A. Peres, *Conserv. Biol.*, 2000, **14**, 240–253.
59. J. G. Robinson and E. L. Bennett, *Animal Conserv.*, 2004, **7**, 397–408.
60. D. Pauly, R. Watson and J. Alder, *Philos. Trans. R. Soc. London, Ser. B*, 2005, **360**, 5–12.
61. A. Balmford, R. E. Green and J. P. W. Scharlemann, *Global Change Biol.*, 2005, **11**, 1594–1605.
62. B. J. Abiodun, J. S. Pal, E. A. Afiesimama, W. J. Gutowski and A. Adedoyin, *Theor. Appl. Climatol.*, 2008, **93**, 245–261.
63. J. Michael Beman, K. R. Arrigo and P. A. Matson, *Nature*, 2005, **434**, 211–214.
64. R. Lal, *Nutr. Cycling Agroecosyst.*, 2004, **70**, 103–116.

65. J. P. W. Scharlemann, A. Balmford and R. E. Green, *Biol. Conserv.*, 2005, **123**, 317–326.
66. K. Norris, *Conserv. Lett.*, 2008, **1**, 2–11.
67. P. F. Donald, R. E. Green and M. F. Heath, *Proc. R. Soc. London, Ser. B*, 2001, **268**, 25–29.
68. P. F. Donald, F. J. Sanderson, I. J. Burfield and F. P. J. van Bommel, *Agric. Ecosys. Environ.*, 2006, **116**, 189–196.
69. J. M. Tylianakis, T. Tscharntke and O. T. Lewis, *Nature*, 2007, **445**, 202–205.
70. A. Wilby and M. B. Thomas, *Ecol. Lett.*, 2002, **5**, 353–360.
71. D. Pauly, J. Alder, E. Bennett, V. Christensen, P. Tyedmers and R. Watson, *Science*, 2003, **302**, 1359–1361.
72. L. A. Morandin and M. L. Winston, *Agriculture Ecosystems & Environment*, 2006, **116**, 289–292.
73. P. Kareiva, S. Watts, R. McDonald and T. Boucher, *Science*, 2007, **316**, 1866–1869.
74. H. M. Tallis and P. Kareiva, *Trends Ecol. Evolut.*, 2006, **21**, 562–568.

Atmospheric Services

JOHN THORNES

ABSTRACT

The atmosphere is a fundamental component of the Earth System and yet its economic and social value to society, as an essential resource, has largely been taken for granted. Terms such as 'weather services', 'meteorological services' and 'climate services' have existed for some time as part of the commercial and public services offered by national and private meteorological providers. These services are primarily based on providing information about the past, present and future state of the atmosphere rather than its intrinsic properties. The new concept of 'atmospheric services', as proposed in this chapter, relates to the inherent set of natural goods and services provided by the atmosphere that enable life, as we know it, to exist and prosper on planet Earth. Twelve basic atmospheric services have been identified (see Table 1) with a Total Economic Value of between 100 and 1000 times the Gross World Product (GWP). Ecosystem Services have been valued at about twice GWP. This analysis shows that the atmosphere is the most precious and valuable of all natural resources in the Earth System. This chapter attempts to justify the valuation of these atmospheric resources and also infers that the atmosphere should be treated as a global commons, and responsibility for its sustainable management should be shared equally amongst all of society. The atmosphere is fragile and at a time of enhanced climate change it requires very careful management and protection. Indeed, a 'Law of the Atmosphere' may be required, especially at a time when there is rising interest in the possible future need for geo-engineering the climate on a global scale.

Issues in Environmental Science and Technology, 30
Ecosystem Services
Edited by R.E. Hester and R.M. Harrison
© Royal Society of Chemistry 2010
Published by the Royal Society of Chemistry, www.rsc.org

1 Introduction: The Atmosphere as Part of the Earth System

The great goddess Athena, the queen of the air;
having supreme power both over its blessings of calm,
and wrath of storm;
and spiritually, she is the queen of the breath of man,
first of the bodily breathing which is life to his blood,
and strength to his arm in battle.[1]

The Earth's atmosphere is arguably the most valuable and at the same time one of the most vulnerable resources on the planet, and yet it has been almost totally taken for granted in the past due to its invisible nature.[2–6] This chapter attempts to compile a critical resource geography of atmospheric goods and services to demonstrate the atmosphere's inexorable value for life on our planet. The atmosphere and its components, weather and climate, are key for our day-to-day survival and well-being.

The atmosphere is a vital part of the Earth System and interacts with the biosphere, the hydrosphere and the lithosphere across the earth's surface:

- The *Lithosphere* contains the soil at the Earth's surface, the solid rock of the Earth's crust, the hot semi-solid rock that lies beneath the crust, the hot liquid rock near the centre of the Earth, and the solid iron core;
- The *Hydrosphere* contains all of the Earth's solid, liquid and gaseous water;
- The *Biosphere* contains all of the Earth's living organisms; and
- The *Atmosphere* contains all of the Earth's air.

These spheres are closely connected and overlap. For example, insects (biosphere) fly through the air (atmosphere), while water (hydrosphere) can flow through the soil (lithosphere), and a change in one sphere is likely to result in a change in one or more of the other spheres. The Earth System is constantly been bombarded by shortwave solar energy (and meteorites) and in turn emits longwave radiation to space. The equilibrium surface temperature of the Earth is currently about 15 °C. Human well-being is wholly dependent on the day-to-day services provided by the Earth System. This chapter is concerned with those services provided by the atmosphere.

Ecosystem services are primarily concerned with the value for human well-being of the biosphere. The value of the atmosphere, hydrosphere and lithosphere has not yet been considered in any detail. Indeed Costanza *et al.*[7] state:

It is trivial to ask what is the value of the atmosphere to humankind, or what is the value of rocks and soil infrastructure as support systems. Their value is infinite in total.

However, Barnes[3] states that:

Commonly inherited gifts of nature provide more (or at least a comparable amount of) wealth to humanity than all human efforts combined . . . A market system that values such an enormous trove of wealth at exactly zero is fundamentally flawed.

The true value of atmospheric services is certainly somewhere between zero and infinity! This chapter attempts to realistically value the individual components of atmospheric services for the first time. In total their replacement value may be effectively infinite (>1000 GWP) but it is certainly not trivial to identify and assess the importance of the atmosphere to humankind at a time when climate change is threatening the very existence of society. To achieve a sustainable atmosphere, which we have to manage effectively, we must understand all its various systems and services to human well-being. Only then can we assess those atmospheric services that need urgent attention and assess what the likely consequences would be to change or adapt them – remembering that all the services are interlinked.

As well as providing vital resources for human well-being, the atmosphere can also be a hazard to society. Atmospheric hazards were estimated by Munich Re[8] to have cost the global economy about £15 billion yr^{-1} in damage costs and 15 500 lives yr^{-1} over the period 1950–2008. The damage costs from Hurricane Katrina alone are estimated to have amounted to about $90 billion.[5] Munich Re also estimate that due to enhanced global warming the annual damage bill is rising and could reach $250 billion by 2050, as can be inferred from Figure 1.

However, these numbers are small in comparison to the estimated impact of enhanced global warming of 'between 5 and 20% of GWP each year, now and forever'[9] which equates to between £2 trillion and £8 trillion. These costs to society of weather and climate hazards on an annual basis are also orders of magnitude less than the benefits of atmospheric services, as will be shown below.

Lovelock[10,11] reminds us through the metaphor of Gaia, that the Earth self-regulates the thin shell of land, ocean and atmosphere so that life can flourish, as it has for the last three billion years. The vitally important part that the atmosphere plays *via* global warming, as part of the Earth System, has been recognised, as society now seeks to mitigate and adapt to climate change. However, the ability of Gaia to self-regulate has been threatened by society:

We have grown in number to the point where our presence is perceptibly disabling the planet like a disease. As in human diseases there are four possible outcomes: destruction of the invading disease organisms; chronic infection; destruction of the host; or symbiosis – a lasting relationship of mutual benefit to the host and the invader.[11]

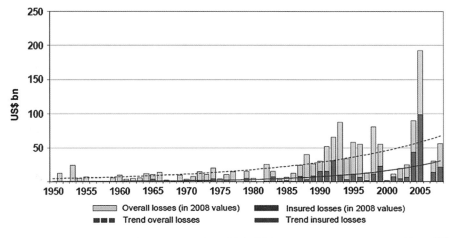

Figure 1 Munich Re estimates of Overall Global Losses due to weather catastrophes.

In order to understand how society can develop the required symbiosis, to prevent disastrous change to our Earth System, we must also understand the basic intrinsic value of the various components of our Earth System to society. Currently there is much research into the value of ecosystem services and their management, but the basic intrinsic value of the atmosphere is, as yet, not understood. The benefits of sustainable management of the atmosphere far outweigh the costs of inaction.

2 Ecosystem Services *versus* Atmospheric Services

The terms 'weather services',[12] 'meteorological services'[13] and 'climate services'[14] have existed for some time as part of the commercial and public services offered by national and private meteorological offices. These services provide *information* about the state of the atmosphere (*e.g.* weather forecasts, climate prediction, weather and climate data sets, weather derivatives, climate emission trading *etc.*). 'Atmospheric services' on the other hand, as proposed in this chapter, refers to the intrinsic set of natural goods and services through which the atmosphere itself enables life on Earth: from the air that we breathe; the protection of the ozone layer; the transmission of sound; the support of air transport; to the provision of global warming. In this chapter twelve basic atmospheric services (there are most certainly more than 12) are explored and their analysis shows that the atmosphere is by far the most precious and valuable of all natural resources in the Earth System – even more valuable than water. This chapter attempts to provide some preliminary values of these atmospheric resources and also argues that the atmosphere should be treated as a global commons. The responsibility for its sustainable management should therefore be shared equally amongst, and governed by, all of society.

The growth in the study of ecosystem services has largely ignored the atmosphere,[7,15,16] with one or two exceptions.[17] Indeed the actual definition and classification of ecosystem services is still widely debated.[18-20] Costanza et al.[7] did include monetary value for the role of ecosystems in the regulation of atmospheric composition, global temperature and precipitation, though few details were given on how these calculations were made. Costanza et al.[7] estimated the global economic value of seventeen ecosystem services to be about $33 trillion per year, compared with the Gross World Product of around $18 trillion per year at that time (approximately twice GWP). Of these seventeen ecosystem services, two are directly related to atmospheric services:

1. Gas regulation, regulation of atmospheric chemical composition (value estimated to be $1.341 trillion); and
2. Climate regulation, regulation of global temperature, precipitation, and other biologically mediated climatic processes at global or local levels (value estimated to be $ 0.684 trillion).

Together these account for just over £2 trillion, or about 6% of the total. Indirectly the atmosphere is also an integral part of several of the other ecosystem services, including water supply, nutrient cycling, pollination, recreation and cultural, which account for more than two thirds of the remaining total. However, Costanza et al.[7] report:

For the purpose of this analysis we grouped ecosystem services into 17 major categories . . . We included only renewable ecosystem services, excluding non-renewable fuels and minerals and the atmosphere.

This is confusing as it is not entirely clear which atmospheric services have been included and which have been excluded. Nevertheless it is clear that the limited atmospheric services that have been included make up a significant proportion of the $33 trillion, and also that many more atmospheric services have been excluded. It is also clear that the value of atmospheric services will be significantly greater than the value of ecosystem services, and that ecosystem services could not exist without the atmosphere.

Whatever the magnitude of the calculations made by Costanza et al.,[7] in the past, ecosystem services, like atmospheric services, have been taken for granted. The Millennium Ecosystem Assessment[21] defines four categories of ecosystem services: *Provisioning Services*, for example ecosystem goods like food, fuel and fresh water; *Regulating Services*, like air quality and climate regulation and water regulation; *Cultural Services*, such as recreation and aesthetic value; and *Supporting Services*, for example the production of oxygen and water cycling by ecosystems. These categories have also been applied to atmospheric services in this chapter as shown in Table 1.

Table 1 The Twelve Atmospheric Services.

Rank in Value	Atmospheric Services	Usage Trend	At Risk	Entity	Service Type
1	The air that we breathe	+ +	**	O_2, N_2 etc	Provisioning
2	Protection from radiation, plasma and meteors	+	**	Density, Ozone Layer	Supporting
3	Natural global warming of 33 degrees Celsius	+	*****	CO_2, CH_4, N_2O, H_2O^{++}	Supporting
4	The cleansing capacity of the atmosphere and dispersion of air pollution	+	*	OH, Wind, Temp	Regulating
5	The redistribution of water services	+	**	H_2O	Supporting
6	Direct use of the atmosphere for eco-systems and agriculture	+	*	CO_2, N_2 Filtered Solar	Provisioning & Supporting
7	Combustion of fuel	–		O_2	Provisioning
8	Direct use of the atmosphere for sound, communications and transport	+	*	Density Pressure	Supporting
9	Direct use of the atmosphere for power	+ +		Wind, Wave	Provisioning
10	The extraction of atmospheric gases	+		O_2, N_2, Ar etc	Provisioning
11	Atmospheric recreation and climate tourism	+	*	Sun, Wind, Clouds, Snow	Cultural
12	Aesthetic, spiritual and sensual properties of the atmosphere, smell and taste	+		Sky, Clouds	Cultural

3   The Atmosphere as both a Resource and a Hazard

A critical resource geography of the atmosphere should attempt to understand the historical and spatial development of the social, political, economic and cultural processes that have led to the atmosphere being used as a service and as a resource.[5,22–26] Firstly, we have to understand the basic properties of the atmosphere in order to appreciate its importance as a basic set of commodities and services. What regulations exist for its use? Who owns the atmosphere? Is the atmosphere a common? How much is the atmosphere worth? What is its value to society? There are many more questions than answers at this stage of enquiry, but this chapter will attempt to fuel the dialogue between physical and social scientists.

We are all familiar with considerations of the atmosphere as a hazard, as it makes the news headlines virtually every day with floods, hurricanes (tropical cyclones), gales, snow, tornadoes and droughts happening somewhere all the time. What is a hazard for one part of society may be a resource for another. For example, tropical cyclones provide a significant proportion (half) of the annual rainfall in Japan and also considerable new business for the construction industry every year.

A physical property of the atmosphere, such as temperature or rainfall, can be considered as a resource within certain thresholds, and as a hazard when outside those thresholds. The value of the thresholds may change with new technology and if the climate changes, the proportion of time the temperature or rainfall stays within the resource thresholds will change. Each economic activity will have different thresholds. For example, if the physical parameter is rail surface temperature, then an upper threshold of 53 °C identifies the potential hazard of rail buckling due to heat in the summer, and 0 °C identifies the potential hazard of points freezing in the winter.[27] The identification and prediction of thresholds using a form of weather sensitivity analysis is still under researched. If one is to identify and manage the potential impact of climate change, then the existing sensitivities and thresholds need to be known.

The atmosphere is very different to most conventional resources, like oil or coal, or services like banking or insurance, or ecosystem services. It covers the entire planet and is constantly being mixed by the weather and *via* the hydrological, carbon, nitrogen and sulfur cycles. The atmosphere is remarkably consistent in its composition around the globe but the variations in solar radiation, due to the Earth's shape, tilt and rotation, ensure that the weather and climate vary significantly across the Earth's surface. Solar radiation is also taken for granted and is both a hazard to life (sunburn and cancer) but is also a huge service for life on Earth. The atmosphere removes most of the ultraviolet radiation *via* the ozone layer to reduce significantly the hazard. The atmosphere lets most of the other wavelengths of solar radiation through, but then traps the outgoing longwave radiation to provide natural global warming for the planet of about 33 °C. For the purpose of this chapter, the filtering of solar radiation will be included as an Atmospheric Service.

There are effectively three types of atmospheric commodity:[5]

1. The material atmosphere itself (*e.g.* oxygen, nitrogen, argon, carbon dioxide, water and spatial dimensions, *i.e.* volume);
2. The physical/chemical properties of the atmosphere (*e.g.* temperature, pressure, density, wind, clouds, precipitation, radiation, optical effects and electrical charge) which comprise the weather and can be directly exploited (*e.g.* for agriculture or power); and
3. Data, information or predictions about the atmosphere used by weather forecast providers, insurers and a host of other services;

The first two can be broadly labelled as atmospheric resources and the third as a new atmospheric paradigm concerned with economic instruments, such as weather derivatives and climate emissions trading, as well as more traditional methods of weather observation, weather forecasting and climate prediction.[5] This chapter will be concerned with both the material, physical and chemical properties of the atmosphere, and the services these properties provide.

4   Who Owns the Atmosphere?

Before we can attempt to put a value on the atmosphere, or discuss the composition and management of atmospheric services, we need to examine who owns the atmosphere.[28] There is an urgent need for society to consider the atmosphere as a precious global entity that requires global management. Currently the atmosphere is managed in a non-sustainable piecemeal way. The artist Amy Balkin has attempted, unsuccessfully so far, to put forward the whole atmosphere as a UNESCO World Heritage Site.[29] There is a need to pull together existing atmospheric regulations relating to airspace, air quality, acid rain, ozone depletion and climate change to establish a 'Law of the Atmosphere' for the global 'Atmospheric Commons'.

Airspace above countries is governed by air traffic control systems for civil air travel, and protected zones for military use are closed off from civilian aircraft use. Payment is made each time an airline enters and leaves the airspace, to cover the costs of air traffic control. But specification of a national volume of air, or the protection of a specific composition of air is not feasible. due to the dynamic nature and movement of the dispersing air. The gaseous phase of the Earth System held within the Earth's atmosphere is therefore governed differently to the liquid phase (hydrosphere, plus liquid fossil fuels), or the solid phase (land), or even the gaseous phase of materials extracted from either the liquid of solid phases (natural gas or gasified fossil fuels).

Does the atmosphere belong to everyone or has it been enclosed by governments or private companies? Ausubel[2] discusses in some detail whether or not the atmosphere can be considered a common and quotes Schauer's[30] four requirements that define a common:

1. A common must exist within and as a part of a wider rule or custom.
2. A common must be identified by practical laws or rules which distinguish it from what is not a common.
3. A common must be open to community or public use and closed to exclusive appropriation.
4. a common must be, by nature or as a result of laws or rules applied thereto, in such a condition that use by some does not preclude or significantly interfere with use by others.

The atmosphere fulfils all four of these postulates in part, but some reservations remain for each one.[2] Barnes[3] outlines the four basic definitions of property rights which are based on Roman Law:

1. *Res privatae*: private things – things in possession of an individual or corporation.
2. *Res publicae*: public things – things owned and set aside for public use by the government, such as public buildings, highways and navigable waterways.
3. *Res Communes*: common things – things accessible to all that can't be exclusively possessed by an individual or government.
4. *Res nullius*: unowned things – things that have no property rights attached until they're taken into possession and become *res privatae* or *res publicae*.

Like Roman law, English law distinguished between two kinds of public property, one belonging to the state and the other to all citizens. The Magna Carta, signed in 1215, established fisheries as a res communes, a commons available to all. Similar status was given to the air, running water and wild animals. There's thus an old and clear distinction between common property and state property, and the air falls decidedly into the common category.[3]

The sky is nothing if not the ultimate commons. We all inhale oxygen from it, exhale carbon dioxide into it, and use it daily in many other ways. On the theory that use implies ownership, or simply that commoners own the commons, the sky should be our common property.[3]

Is Barnes right? Buck[31] in her book *The Global Commons* concludes that:

The physical characteristics of atmospheric resources (clean air and stratospheric ozone) are not analogous to resources in common domains, and the atmosphere regime therefore is not a commons regime. Public policy analysis that approaches atmospheric pollution as a negative externality controlled by protective regulatory policy provides a better approach to the problems of the regime and their analysis.

However her analysis is restricted to considerations of polluting the atmosphere, rather than regulating the exploitation of the material atmosphere.

Barnes[3] further suggests the setting up of a 'sky trust' in the United States, similar to the Alaska Permanent Fund, that would operate a 'Cap and Dividend' system for carbon emissions. What is important about his scheme is that it would require tradable carbon emission permits at *source*. Each company that extracts fossil fuels would need to buy the right to emit each resultant tonne of carbon each year. The total number of permits would be reduced each year and if a company exceeded their allowance they would have to buy permits from companies that had spare permits, or pay a fine. The money raised would be paid into a mutual trust to be shared equally amongst all citizens.

Tickell[32] treats the atmosphere as a global commons whilst recognising that air pollution is a kind of negative commons. As Hardin[33] states:

In a reverse way, the tragedy of the commons reappears in problems of pollution. Here it is not a question of taking something out of the commons, but of putting something in – sewage, or chemical, radioactive, and heat wastes into water; noxious and dangerous fumes into the air, and distracting and unpleasant advertising signs into the line of sight . . . The rational man finds that his share of the cost of the wastes he discharges into the commons is less than the cost of purifying his wastes before releasing them. Since this is true for everyone, we are locked into a system of 'fouling our own nest', so long as we behave only as independent, rational, free-enterprises.

Tickell[32] describes the advantages and disadvantages of the various solutions that have been put forward to achieve a low carbon economy that also recognise the atmosphere as a global commons: Contraction and Convergence, Cap and Share, Cap and Dividend, and Carbon Rationing with Variable Quotas. These schemes are concerned with stabilising greenhouse gas concentrations in the atmosphere at a level which has been targeted at 350 ppm CO_2 equivalent. Tickell[32] suggests a similar solution to Barnes[3] that would regulate greenhouse gas emissions 'upstream' at the oil refinery, the coal washing station, the gas pipeline and the cement factory *etc.* by the auctioning of global permits. The proceeds, estimated to be of the order of 1 trillion Euros per year, would be placed into a Climate Change Fund which would be used to finance mitigation and adaptation measures around the globe.

Successful intervention strategies to slow or prevent pollutant emissions to reverse environmental degradation are numerous, and sometimes on a global scale. The Montreal Protocol has successfully used a direct regulation global approach to phase out ozone depleting chemicals using a 'Multilateral Fund' to help developing countries achieve targets. Indeed, because the controlled gases are also very effective greenhouse gases, the Montreal Protocol has been significantly more successful than the Kyoto Protocol in reducing radiative forcing. However, these policies tend to consider trace elements that are man-made, or at concentrations which are orders of magnitude higher than those normally found in the environment. It is perhaps the ubiquity of carbon in our environment that creates difficulty for legislators to delineate forms of carbon.

This relatively easy and rapid success of the Montreal Protocol in tackling (ozone depleting) greenhouse gas emissions stands in stark contrast to the slow, meagre and expensive gains achieved under the Kyoto Protocol. This strongly suggests that there is a role for direct regulation, also backed by a 'Multilateral Fund' or similar instrument, in a future climate protocol.[32]

The Montreal Protocol is on course to reverse the 'hole in the ozone layer' and therefore at least one example of 'fouling our own nest' through the design of 'material culture', in this case mostly refrigerators, has been solved by global regulation.

Hulme[26] argues that:

. . . climate change possesses all the attributes of a 'wicked' problem, a situation defined by uncertainty; inconsistent and ill-defined needs, preferences and values; unclear understanding of the means, consequences or cumulative impacts of collective actions; and fluid participation in which multiple, partisan participants vary in the amount of resources they invest in resolving problems. 'Tame' problems, on the other hand – while they may be complicated – have relatively well-defined and achievable end-states and hence are potentially solvable. The example of stratospheric ozone depletion may fall into this category.

Thornes and Randalls[5] discuss how 'biopower' regulation had been replaced by neo-liberalism through the use of economic instruments set up as part of the Kyoto Protocol. The current credit crunch suggests that a return to biopower style regulation is overdue and that the Tickell approach would provide a more effective solution for a new Kyoto Protocol – as discussed in December 2009 in Copenhagen for implementation in 2012. The atmosphere has already been commodified in certain respects, however, and the acceptance of the atmosphere as a global commons needs to be agreed as soon as possible. In order to achieve this we need to recognise which parts of the atmosphere have been commodified and which parts can still be regarded as a common. What is needed is a United Nations Convention on the 'Law of the Atmosphere' similar in scope and regulation to the 'Law of the Sea' which was ratified by the EU in 1998. The UN Convention on the Law of the Sea defines the pollution of the marine environment as:

. . . the introduction by man, directly or indirectly, of substances or energy into the marine environment, including estuaries, which results or is likely to result in such deleterious effects as harm to living resources and marine life, hazards to human health, hindrance to marine activities, including fishing and other legitimate uses of the sea, impairment of quality for use of sea water and reduction of amenities.

A similar Law of the Atmosphere might define pollution of the atmospheric environment as:

> *. . . the introduction by society, directly or indirectly, of substances or energy into the atmospheric environment, which results or is likely to result in such deleterious effects as harm to living resources, hazards to human health, hindrance to atmospheric services, impairment of air quality and reduction of amenities.*

Any Law of the Atmosphere would also need to contain conventions with regard to regulating the extraction of substances and energy from the atmosphere and also regulating its use by private individuals, companies, organisations and governments. Thus a Law of the Atmosphere must control: 1. putting material things into the atmosphere (*e.g.* air pollution); 2. taking material things out of the atmosphere (*e.g.* commodities); and 3. using the atmosphere (*e.g.* airspace, noise pollution).

5   The Valuation of Atmospheric Services

Trying to put a value on nature is extremely controversial and difficult. There are a host of papers discussing the basic principles and pitfalls.[34–36] The Earth System and solar energy provide all the natural capital and services that underlie human well-being and GWP. The relative contributions of the atmosphere, biosphere, hydrosphere, lithosphere and solar energy to GWP are impossible to differentiate and value accurately. Recent estimates of the value of a statistical human life in developed countries include £4.6 million (US EPA) and £1.4 million (UK DfT).[60]

The controversy about the 'value of a statistical life' in the context of climate change economics was particularly heated in the 1996 Second Assessment Report of the IPCC. Using conventional neoclassical economic procedures it was suggested that the value of 'statistical life' in the developed world (∼ $5m) might be at least ten times higher than in the developing world (∼ $0.5m).[26]

In attempting to assess the economic value of atmospheric services let us consider a different approach. The 2008 Gross World Product GWP was approximately $70 trillion (exchange rate £1 = $1.6) *i.e.* £43 trillion[37] (for simplicity let us give it the value of 1 GWP). If we had to replace the services provided by the atmosphere (for example by using some form of geo-engineering or by moving the Earth's population to another planet) the total economic costs would certainly be multiple times the current value of GWP. For example, the current modest estimates of the costs of sulfate aerosol injection into the stratosphere only consider direct costs[46] and have not considered the huge indirect and other costs. If the atmosphere suddenly stopped delivering services then the entire human race would be wiped out in a few minutes.

Although the use of GWP for assessing the social benefit of scarce resources has been criticised by many authors[26,38] as being too 'materialistic' or

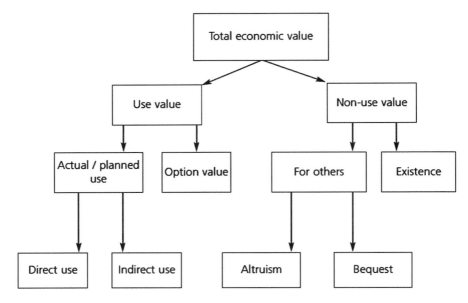

Figure 2   Total Economic Value.

'completely inadequate' it nevertheless does provide a convenient 'yardstick' for comparison with Costanza *et al.*[7] and Stern.[39] A useful summary of the various approaches and methodologies that have been undertaken in trying to evaluate ecosystem services has recently been published by the Department for Environment, Food and Rural Affairs.[40] The concept of Total Economic Value (TEV) is summarised in Figure 2.

TEV takes into account both the use and non-use values individuals and society gain or lose from marginal changes in ecosystem services.[40]

Note that this suggests the analysis of *marginal changes* in ecosystem services which obviously links to policy:

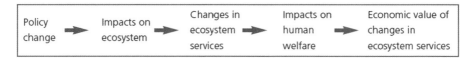

For atmospheric services we are also interested in their 'intrinsic value' and from an economic point of view we need to concentrate on likely changes to atmospheric services and the 'opportunity costs' involved. For example:

In assessing the value of clean air, we will want to measure the amount of other resources society would be willing to devote to improving the quality of air (or preventing its degradation).[38]

This concentration by economists on small marginal changes means that they are usually additive and clear cut. If non-marginal changes are valued then it becomes more complicated:[38]

Suppose we were to ask what is the 'total' value of atmospheric services. Since we would not survive if all these services were removed, this total must be more or less, all the other resources we have. By the same token the total value of all water services is also pretty much everything. Consequently measures cannot be additive since the total value (opportunity cost) of atmospheric and water services cannot exceed all the resources we have to pay with. The problem here is that once atmospheric services are gone, water resources no longer have value and vice versa.

Interestingly Starrett[38] continues in a footnote:

It is because nonadditivity was ignored that Costanza et al. (1997) derived a number for the value of ecosystem services that exceeded the world GWP (a proxy for total ability to pay). I would argue that their aggregate number is meaningless although the component estimates might still be useful in some contexts.

Thus economists, from a practical policy point of view, prefer to operate within the margins of a single GWP. However, as Starrett[38] admits, the component estimates are useful to raise awareness of the individual importance of all environmental services.

Kramer[41] suggests for simplicity the formula:

Total Economic Value = Use Value + Indirect Use Value + Option Value + Nonuse Value

The various methodologies for evaluating Total Economic Value (Defra 2007)[40] are summarised in Figure 3.

A variety of methodologies has been used in the past to estimate TEV by assessing an individual's or a community's Willingness to Pay (WTP) for improvements and/or Willingness to Accept (WTA) compensation for change. Hyslop,[42] for example, looks at visibility and what people are willing to pay (WTP) for an increase in visibility or to accept payment (WTA) for a decrease in visibility. Another study[43] found that a reduction of $1\,\mu g\,m^{-3}$ in PM (particulate matter) was worth between \$21 and \$337 per person per year.

5.1 An Estimate of the Total Economic Value of Atmospheric Services

The value of atmospheric services is obviously neither zero nor infinity as discussed in the introduction. Carbon dioxide already has a virtual market value of about £10 to £12 per tonne in the EUETS (European Union Emissions Trading Scheme).[i]

---

[i] Note that it is recognised that this is a heuristic approach and is likely problematic given that the EU's prices are based on the distribution of an artificially scarce resource, whereas all atmospheric carbon is not scarce.

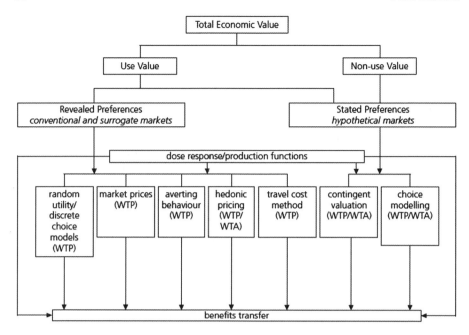

Figure 3   Methodologies for assessing Total Economic Value.

In total there are approximately 3 trillion tonnes of carbon dioxide in the atmosphere which therefore has a virtual value of about £30 trillion. If the rest of the atmosphere were valued at £10 per tonne, the 5148 trillion tonnes of atmosphere would be worth more than 1000 GWP. Table 1 shows that every tonne of air is providing services for human well-being whether that tonne of air is near the surface or in the upper reaches of the atmosphere.

A tonne of air does not really relate to our every day experience. The more common measure of air is the cubic metre (the density of air is around 1.2 kg per cubic metre at sea level). The current market price of compressed air is about £2 per cubic metre (see section 6.1), although the price is much reduced for bulk purchases. What would we be 'willing to pay' for our use of the atmosphere? Let us make a conservative estimate that the atmosphere has a TEV (Total Economic Value) of somewhere between 0.1 p and 1 p per cubic metre for all the atmospheric services listed in Table 1. Each person on the planet breathes about 5500 cubic metres of air per year and this would give a value of between £5.50 and £55 per year. Certainly people, even in developing countries, would recognise that £5.50 per year (1.5 p per day) for the air that they breathe, plus all the other services provided in Table 1, is a bargain. Even at £55 per year (15 p per day) this would be very reasonable compared to the cost of water and land rates in developed countries. If we therefore value the atmosphere at between 0.1 p and 1 p per cubic metre, the whole atmosphere (4.3×10^{18} cubic metres) would be worth ($£4.3 \times 10^{15-16}$), which is the equivalent of between 100 to 1000 GWP.

The Total Economic Value (TEV) of the twelve atmospheric services item-ised in Table 1 is therefore estimated by two different methods to be at least somewhere between 100 and 1000 GWP, which is a very low estimate as it only includes direct 'use value' by humans for atmospheric services.

Table 1 shows a ranking value for each atmospheric service. These figures are initial estimates and are only meant to give a broad idea of the importance of each service for human well-being. An interesting question to ask is how much would it cost to replace each these vital atmospheric services using, for example, some form of geo-engineering? Boyd[44] ranks the various geo-engineering schemes that have been proposed to mitigate climate change, but cautions that many of them have unwanted and costly side-effects. Cathcart and Æirkoviæ[45] describe enclosing the entire Earth's atmosphere with a 'polyvalent roof' so that weather and climate become an 'air-conditioning' issue! The Royal Society have just published the report 'Geo-engineering the Climate'[46] which looks at the current ideas for extracting carbon from the atmosphere and reducing the amount of solar radiation reaching the Earth's surface. None of these ideas have yet been effectively costed,[47] but a law of the atmosphere would need to be in place before any global geo-engineering experiments could begin.

The ranking in value of each atmospheric service in Table 1 is subjective, but very important, as those services at risk (1–6, 8 and 11) include the top six in terms of economic value and importance to human well-being. The rating of the 'at risk' services is also subjective and needs further research.

6 Atmospheric Services and Natural Capital

In this section, we examine each atmospheric service in turn to assess their natural capital and contribution to market value.

Air has a material value. You can buy air in shops, over the phone or on the internet and have it delivered to your premises. Normally it will be compressed into canisters or cylinders and be known as 'compressed air'. Compressed air has a myriad of uses and it has been estimated that in Europe nearly 10% of electricity (80 terawatt hours) used by industry is used to produce compressed air. Scuba divers obviously need air underwater – the word scuba stands for 'self-contained breathing apparatus'. Today you can pay more than £2 per cubic metre for compressed air (see Table 2).

Table 2 Gas prices per cubic meter supplied by BOC (as of March 2009).

Gas	Price
Compressed Air	£2.32
Oxygen	£1.84
Nitrogen	£2.38
Hydrogen	£5.98

6.1   The Air that We Breathe

Clean air is considered by the World Health Organisation[53] to be a basic requirement of human health and well-being. The fundamental act of breathing by humans, animals and insects is something that we automatically take for granted.[24] Humans on average breathe about $15\,m^3$ of air a day,[25] converting oxygen into carbon dioxide and water vapour – the two most important greenhouse gases. An adult human exhales about 3.6% carbon dioxide (see Table 3) which amounts to approximately $0.54\,m^3$ carbon dioxide per day. The density of carbon dioxide is $1.98\,kg\,m^{-3}$, so that each adult produces more than a third of a tonne of carbon dioxide per year. The Earth's population is 6.8 billion, so that in a year humans produce about 2.6 billion tonnes of carbon dioxide which represents nearly three quarters of a billion tonnes of carbon $(2.6 \times 12/44)$.

As I contemplate the blue of the sky, I am not 'set over against it' as an acosmic subject . . . I am the sky itself as it is drawn together and unified, and as it begins to exist for itself; my consciousness is saturated with this limitless blue.[48]

Indeed we are the sky – from the moment we take our first breath when we are born, we are the sky. We breathe the sky because the sky is just a light show created by the atmosphere – it is one and the same thing. It is blue during the daylight hours and invisible at night unless obscured by clouds or pollution. Breathing is a performance that is largely invisible to us, but as a consequence we are all insiders to climate change and enhanced global warming. The oxygen from the atmosphere (see Table 3) is actively taken up by our blood to be delivered to our 15 million million cells. Without it we can only last a few minutes before death.

Without oxygen, life on Earth would never have got beyond a slime in the oceans, and the Earth would probably have ended its days in the sterility of Mars or Venus. With oxygen, life has flourished in all its wonderful variety: animals, plants, sex, sexes, consciousness itself. With it, too, came the evolution of ageing and death.[49]

As atmospheric levels of carbon dioxide have been observed to be increasing by about 2 ppmv per year, in recent years, so levels of oxygen have been observed to be decreasing by about 3 ppmv per year.[50]

Table 3   The content of inhaled and exhaled air.

Medium	*Inhaled Air*	*Exhaled Air*
Nitrogen (incl. rare gases)	78.62%	74.9%
Oxygen	20.85%	15.3%
Carbon dioxide	0.038%	3.6%
Water vapour	0.492%	6.2%

Nitrogen is also important to life on Earth – it is one of the essential building blocks for proteins in our bodies and why we need to eat vegetables, either legumes that can 'fix' nitrogen directly from the air, or other vegetables that acquire nitrogen from the soil. Nitrogen also plays a vital role in diluting the presence of oxygen, otherwise the atmosphere would burst spontaneously into flame.

We don't just live in the air. We live because of it.[6]

6.2 Protection from Extra-Terrestrial Radiation Plasma and Meteors

Five billion tonnes of ozone in the stratosphere protect us from ultraviolet radiation from the sun.[6] Not that all ultraviolet radiation is bad for us.

UVA (400–320 nm) stimulates the manufacture of Vitamin D in our bodies, as well as triggering the production of melanin by our skin cells to give a protective suntan (for people with white skin). Thankfully it manages to get through the ozone layer. UVB (320–290 nm) is dangerous and although much is absorbed by the ozone layer, what does get through can course sunburn, skin cancer and cataracts. It also inhibits photosynthesis and reduces plant growth. UVC (290–100 nm) would be the most damaging ultraviolet radiation, but it is totally absorbed by the ozone layer.

The hole (>60% depletion) in the ozone layer over the Antarctic was first identified in 1985. By 2000, it covered an area of 28.3 million km^2, but since that time, due to the Montreal Protocol, there are signs of recovery. Chlorofluorocarbons (CFCs) which destroy the ozone were banned and global levels are gradually now falling.

The atmosphere, in conjunction with the Earth's magnetic field, also protects us from the solar wind (radiation plasma). The flickering lights of the aurora borealis show when the ionosphere is absorbing radiation plasma from the sun. The Earth's magnetic field captures these streams of electrons and directs them towards the Poles where the atmosphere soaks them up and glows with their energy.

Another spectacular light show is caused by meteors disintegrating as they enter the atmosphere.

The upper air burst into life!
And a hundred fire-flags sheen,
To and fro they were hurried about!
And to and fro, and in and out,
The wan stars danced between.[66]

Extraterrestrial material entering the Earth's atmosphere produces heat and light as it plummets to the Earth's surface. Meteor showers occur all around the world, normally burning up in the atmosphere before impact. Meteorites reaching the ground account for about 100 tonnes of material at the Earth's surface each year. If it were not for the atmosphere, the Earth's surface would be covered in craters like the moon.

6.3 Natural Global Warming

Natural global warming keeps the Earth's surface at an average temperature of about 15 °C. Without the atmosphere the equilibrium, temperature of the Earth would be about −18 °C, hence natural global warming is responsible for an increase in temperature of 33 °C. Enhanced global warming caused by society polluting the atmosphere so far amounts to an additional 0.8 °C, but we are already committed to about 1.5 °C due to extra energy stored in the oceans and not yet released. The IPCC has estimated that an effective doubling of greenhouse gases could raise the equilibrium surface temperature by 3 °C by 2050. Thus our atmosphere is vital to sustain life on the planet as we know it and we need to act quickly to prevent the atmosphere from overheating our planet.[9,67] The redistribution of energy from the tropics to the poles fuels our weather and creates the current distribution of climate around the globe.

6.4 Cleansing Capacity and the Dispersion of Air Pollution

The atmosphere has evolved over billions of years and remains dynamic. The first atmosphere, just after the Earth was formed, was composed mostly of hydrogen and helium which escaped into space. The second was formed around 4.5 billion years ago as the Earth solidified and out-gassing created an atmosphere of carbon dioxide, nitrogen, sulfur dioxide and water vapour. Then about 2.5 to 3 billion years ago cyanobacteria first started to photosynthesize and convert carbon dioxide into oxygen. Today the atmospheric trace gas composition is regulated by the oxidation capacity of the atmosphere:

The atmosphere is a chemically complex and dynamic system that interacts significantly with the land, oceans, and ecosystems. Most trace gases emitted into the atmosphere are removed by oxidizing chemical reactions involving ozone and the hydroxyl free radical. The rate of this self-cleansing process is often referred to as the oxidation capacity of the atmosphere. Without this process, atmospheric composition and climate would be very different from what we observe today.[52]

The atmosphere has always been used as a waste dump by society and as a consequence, as Hardin[32] stated, we are fouling our own nest. There are other perils caused by atmospheric dispersion as many infectious diseases are spread through the air. SARS (Severe Acute Respiratory Syndrome) infected more than 8000 people and killed 813 in 2003. The common cold, diphtheria, influenza, measles and mumps are passed on through the air, primarily *via* coughing and sneezing.

A single sneeze can propel an aerosol of 2 million virus particles a distance of nine metres. On average it takes just ten particles to establish the disease in a new host; in theory, one sneeze, properly disseminated, could infect a small city.[4]

The atmosphere is not pristine. The air is full of small particles of spores, pollen, viruses, dust, algae and molds – typically in urban areas we inhale about 25 million particles with every breath. At any one time there can be up to 3 billion tonnes of particles in the atmosphere. Many of these are natural resulting from dust storms, volcanic eruptions, sea spray and lightning strikes causing forest fires. Human activities add to this naturally varying atmospheric pollutant load, and for individual pollutants may increase concentrations by several orders of magnitude. This is most apparent in the cases of plasticizers (phthalates) and nuclear fall-out (strontium). Anthropogenic air pollution from factories, chemical works, vehicle exhausts, power stations, and domestic sources is least dispersed close to the source (for example, in cities or industrial zones), and lowest levels of air pollution are found over the oceans and Antarctica where effective dispersion has taken place. However, there is extensive evidence that background levels of pollutants such as oxides of nitrogen, are increasing in the homogenised system. Because of this, emissions of air pollutants such as sulfur dioxide and particles have been increasingly regulated at source across the world since the 1960s. Ambient outdoor levels of selected pollutants have therefore also been regulated to great effect in Europe and the USA to increase local responsibility for air quality. The controlled pollutants are all generated or linked indirectly to fossil fuel combustion. Since the seminal epidemiology paper by Dockery *et al.*,[54] hundreds of international studies on human exposures to increased levels of air pollution have demonstrated links, even when other confounding factors are taken into consideration. Toxicology has similarly linked air pollution doses to health effects. A catalogue of human health effects, ranging from cardiovascular to infectious disease, are reported and many are linked directly to combustion gas release.[55] Children growing up in areas of higher air pollution experience lifelong effects in lung function.[56] Since we tend to spend 90% of our time indoors and indoor pollution levels have a smaller dilution capacity, indoor air quality is also regulated and indoor smoking bans have been introduced to prevent exposures to environmental tobacco smoke (ETS), the most important indoor air pollutant. Exposure to wood smoke from fires for cooking and heating is also an important problem in developing countries. The cost of air pollution is routinely calculated by policy makers in the UK and each time it is examined there is more pressure to reduce both ambient and emissions air quality standards. The more we understand health effects, the bigger the effect appears to be.

The air quality improvements in the developed world are inverted in rapidly developing and undeveloped countries. Air quality is quickly deteriorating in rapidly developing countries, as pollution shifts from the developed to the developing world. Poor indoor air quality remains one of the key health challenges for these countries[57] and international guideline levels of particulates, nitric oxides, ozone and sulfur dioxide are exceeded particularly in developing countries such that:

Air pollution is a major environmental risk to health and is estimated to cause approximately 2 million premature deaths worldwide per year.[80]

If it wasn't for the atmosphere dispersing air pollutants and we all ended up breathing neat exhaust and stack gases, then hundreds of millions of people would die each year!

6.5   Clouds and the Hydrological Cycle

The total volume of water in the hydrosphere (including the oceans, seas, rivers, lakes, ground water, ice, clouds and water vapour in the air) is about 1.4×10^{19} cubic metres, which is about a third of the volume of the atmosphere. Almost 97% of the water in the hydrosphere is undrinkable seawater and another 2% is locked up in the polar ice caps, leaving just 1% for human use and well-being. As the world population grows water supplies are becoming scarcer and increasingly commodified as 'blue gold'. Global fresh water use is expected to rise by up to 40% by 2020. The hydrological cycle plays a key role in redistributing this fresh water around the world. Approximately 5×10^{11} tonnes of water falls in the form of precipitation per year and an equal amount of water evaporates to retain a global water balance. A typical water molecule only spends about nine days at a time in the atmosphere. If fresh water costs about £1 per cubic metre (UK price, 2009) then global precipitation would be valued at about $£5 \times 10^{11}$ which is £500 billion. In other parts of the world, water prices are much higher. Global sales of bottled water are estimated to be $22 billion at an average price of about 70p per litre = £700 per cubic metre. The value of fresh water will rise significantly in some countries over the next few years as climate change melts glaciers and redistributes precipitation. Hoffman[51] estimates that the global cost of clean water through to 2025 will be close to $1 trillion per year.

The hydrological cycle is a natural process that uses the atmosphere to transport water around the globe. When water evaporates from the sea it is transformed into fresh water and then transported by the atmosphere to fall as precipitation elsewhere. The spatial distribution of rainfall is very uneven and some areas receive more rainfall than required and suffer floods, whereas other areas receive less rainfall than required and suffer droughts. Some areas suffer both at different times of the year. Weather modification attempts to either increase or decrease precipitation depending upon the atmospheric conditions at the time. Globally approximately 70% of available fresh water is used by agriculture to grow food and increasingly biofuels.

Clouds provide a vital role in the atmosphere. Cloud formation *via* condensation and sublimation releases huge amounts of energy to the atmosphere to fuel our weather. Clouds host the microphysics that turns cloud droplets into precipitation, and clouds help the wind to redistribute fresh water around the planet. Clouds also control the radiation balance of the atmospheric system and the representation of clouds is still a major uncertainty in weather fore-casting and global climate models. Clouds also present vital visual clues for the art of weather forecasting and some would say clouds and cloudscapes are the 'greatest show on Earth'. The cultural value of clouds is discussed in section 6.12.

6.6   Direct Use of the Atmosphere for Ecosystems and Agriculture

The natural carbon dioxide and oxygen in the atmosphere provide vital ingredients for photosynthesis and respiration. All of the crops that we grow need to fix carbon dioxide, and this has a big global impact on the carbon dioxide levels. Carbon dioxide is an extremely valuable resource therefore, and increases in the amount of carbon dioxide in the atmosphere have led to increased productivity particularly in the northern hemisphere. It is common practice for farmers to increase productivity in greenhouses by pumping in additional carbon dioxide.

From an economic point of view, climate is matter and energy organised in a certain way. If a climatologist were to say to a farmer that the climate is going to change, the farmer could interpret this to mean that deliveries of matter and energy may be going to change in quantity, time, and place, in ways similar to how supplies of fertilizer or gasoline might change.[2]

Almost 100 million tonnes of nitrogen fertilizer (mostly ammonia, ammonium nitrate and urea) are produced each year using nitrogen from the air combined with hydrogen from natural gas. This inorganic fertilizer sustains as many as two out of every five people alive today.[58] Between 1% and 2% of energy production worldwide is for the production of nitrogen-based fertilizer using the Haber process. At an average cost of about £200 tonne^{-1}, this means that global agriculture pays about £20 billion per year for nitrogen fertilizer.[59] The CIA World Fact Book[37] estimated that agriculture provides 4% of global GWP. This represents a turnover of about £2 trillion. Without the filtering of solar radia-tion, precipitation, nitrogen fertilizers, natural global warming and pollination provided by the atmosphere – agriculture would not be possible.

The global winds are also responsible for regulating the upwelling of nutrients for the marine biosphere, as well as contributing (with rain and frost) to the erosion of the Earth's crust thereby replenishing soils and the supply of metallic ions needed to sustain life.

Through their effect in mediating the geographical distribution of upwelling and the depth of the mixed layer, year-to-year changes in the atmospheric circulation, such as those that occur in association with El Niño, perturb the entire food chain that supports marine animals, seabirds, and commercial fisheries.[50]

6.7   The Combustion of Fuel

The burning of fossil fuels and wood to produce energy and electricity requires oxygen from the atmosphere for combustion. We rely on oxygen being available continuously when we switch on our central heating as well as drive our cars or fly in an aeroplane. For example about 16 kg of air is required to burn 1 kg of petrol. In other words about 2 cubic metres of oxygen are needed to burn 1 litre of petrol. This oxygen combines with the carbon in the petrol to produce carbon dioxide such that 1 kg of petrol will produce about 2.3 kg of carbon dioxide. The typical combustion of petrol also releases large amounts of water vapour and energy:

$$2C_8H_{18} + 25O_2 = 16CO_2 + 18H_2O + 34.8\,\text{MJ}\,\text{litre}^{-1}\text{ petrol}$$

The energy released in this exchange is equivalent to about 10 kilowatt hours per litre of petrol and nearly 11 kilowatt hours per litre of diesel. Hence oxygen from the atmosphere is crucial for the production of energy. We currently utilise oxygen without cost or consideration of value, and with minimal consideration of the cost of the emissions of combustion gases.

Each day globally (see Figure 4) we consume about 1 litre of oil per day per person (6.3 billion litres) and oil consumption provides more than one third of

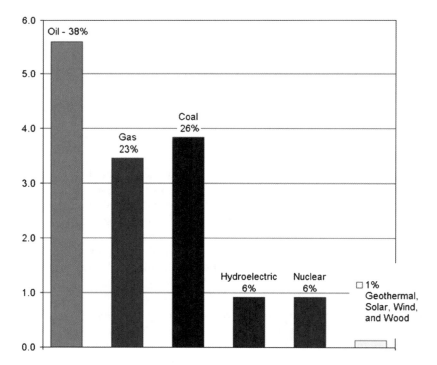

Figure 4   Global Energy Usage in 2004 in terawatts, TW (Total 15 TW).

our daily energy consumption. This means that energy production consumes about the same amount of oxygen each day as we breathe.

6.8 Air Transport, Communications and Sound

Air transport relies directly on the atmosphere. Civil aviation grossed more than $500 billion in 2007 (see Table 4).

Military aviation probably turns over a similar amount, although delineation of civil and military aviation is difficult. With the growth in aviation, the spatial dimension of the atmosphere – airspace – has been increasingly enclosed in order to regulate the flight paths of aircraft for safety and political reasons. Currently civil aircraft (passenger and freight) pay taxes to land and take off at airports but no charge is made for the use of airspace. Aircraft not only use the atmosphere for lift to fly but also freely use oxygen in the atmosphere to combust the aircraft fuel. Aircraft also pollute the lower and upper atmosphere and if the weather conditions are suitable, produce jet contrails which can impact on the local energy balance. Gössling and Upham[61] have discussed the likely impacts of aviation on climate change and the atmosphere. By 2006 there were nearly 20 500 civil aircraft in service worldwide using about 233 million tonnes of fuel per year, equivalent to 733 million tonnes of CO_2 per year. Taking into account the total impact of aviation on climate change, the International Air Transport Association (IATA)[62] estimate that aviation is responsible for about 3% of global anthropogenic greenhouse gas emissions. In view of the projected growth in air travel in the 21st century, aviation has been included in the latest EU emissions trading scheme:

The EU emissions trading scheme entered into force on 2.2.2009, requiring all airlines landing or taking off in EU member states to pay for their carbon dioxide usage through CO_2 allowances or carbon credits beginning in 2012. The target set by the EU is for aviation to reduce greenhouse gases to 3% below the average of 2004–2006 levels in 2012, increasing to 5% for the 2013–2020 period. Estimates of the cost to international airlines in 2013 are around the $2.5 billion mark increasing to $3.5 billion by 2015.[63]

It is hoped that a global carbon emissions scheme will be implemented in the near future, so that aviation emissions can be reduced worldwide in real terms.

Airspace is also required by birds and insects, for travel and sustenance. Birds are physically adapted to flying with hearts up to five times proportionally larger than humans, and avian respiration demands 20% of the body volume compared to 5% in mammals. The large volume of air inhaled supplies copious amounts of oxygen to the chest muscles to sustain the necessary wing flapping for flight. The smaller the bird, the faster the wings have to flap to maintain lift. Insects have to do even more flapping which creates the familiar buzz of the bumblebee (130 wingbeats per second) and the tortuous sound of the mosquito (600 wingbeats per second). A butterfly can

Table 4 Global turnover of civil aviation 2000–2009.

System-wide global commercial aviation	2000	2001	2002	2003	2004	2005	2006	2007	2008F	2009F
REVENUES, $ billion	329	307	306	322	379	413	465	508	536	501
Passenger	256	239	230	249	294	323	365	401	425	394
Cargo	40	39	38	40	47	48	53	58	59	54
Traffic volumes										
Passenger growth, tkp, %	8.6	-2.7	1.0	2.3	14.9	7.0	5.9	5.9	2.0	-3.0
Passenger numbers, *millions*	1672	1640	1639	1691	1888	2022	2124	2260	2304	2236
Cargo growth, tkp. %	9.1	-6.0	8.7	3.9	7.9	0.4	3.9	4.0	-1.5	-5.0
Freight tonnes, millions	30.4	28.8	31.4	33.5	36.7	37.6	39.8	41.6	41.0	38.9
World economic growth, %	4.5	2.2	2.7	2.8	4.2	3.4	4.0	3.8	2.6	0.9
Yield growth, %	-1.1	-2.8	-3.7	2.3	4.5	4.0	6.9	3.6	4.6	-3.0
Yield growth, inflation/ex rate adjusted %	-2.3	-2.9	-5.9	-5.3	-2.0	0.1	3.3	-1.3	0.2	-5.3
EXPENSES, $ billion	318	319	311	323	376	409	450	488	535	497
Fuel	46	43	40	44	65	91	107	136	174	142
% of expenses	14	13	13	14	17	22	24	28	32	29
Crude oil price. Brent, $/b	28.8	24.7	25.1	28.8	38.3	54.5	65.1	73.0	100.0	60.0
Non-Fuel	272	276	270	279	311	318	343	353	361	356
cents per atk (non-fuel unit cost)	39.2	39.7	38.8	38.9	39.5	38.7	40.1	39.2	39.4	39.7
% change	-2.3	14	-2.3	0.3	1.4	-2.1	3.6	-2.1	0.4	1.0
% change, adjusted for ex rate	-0.2	4.1	-3.0	-5.0	-2.4	-2.7	3.3	-4.1	0.6	1.0
Break-even weight load factor, %	60.8	61.5	63.2	62.3	63.4	63.3	63.4	62.8	63.5	62.5
Weight load factor achieved, %	61.5	59.0	60.9	60.8	62.5	62.6	63.3	63.6	62.9	62.2
OPERATING PROFIT, $ billion	10.7	-11.8	-4.8	-1.4	3.3	4.3	15.0	19.7	1.1	3.9
% margin	3.3	-3.8	-1.6	-0.4	0.9	1.0	3.2	3.9	0.2	0.8
NET PROFIT, $ billion	3.7	-13.0	-11.3	-7.5	-5.6	-4.1	-0.1	12.9	-5.0	-2.5
% margin	1.1	-4.2	-3.7	-2.3	-1.5	-1.0	0.0	2.5	-0.9	-0.5

(Source: http://www.iata.org/economics.[62] Note: 2008F & 2009F are Forecasts).

remain airborne with just 10 wingbeats per second by part gliding and part flapping. Some insects, like the snow-dwelling spiders that survive at 6600 metres up the side of Mount Everest are literally fed by the atmosphere in a unique ecosystem. The wind, blowing up the side of the mountain brings a regular supply of algae and insects from below, which was labelled the aeolian biome by the biologist Lawrence Swan in the 1950s.

Plants that are pollinated by animals, particularly birds and bats, are called 'zoophilous', but even they are dependent upon the air. Firstly, they require air for the creatures to fly to them, and secondly for the dispersion of the tantalising odours exuded to attract the pollinators, especially the bats. Insect pollination, or 'entomophily', by bees, wasps, ants, beetles, moths, butterflies and flies is also encouraged by strong scents from the flowers.

Perhaps one of the most fundamental and undervalued services provided by the atmosphere is the dispersion of pollen:

Only about 10% of plants are wind-pollinated but reproductively speaking, these are wildly successful, comprising over 90% of the Earth's total plant population. Such anemophilous or 'wind-loving' plants – the bane of hay-fever sufferers – collectively coat every square meter of the planet each year with 100 million grains of pollen.[4]

All the world's grasses including essential cereal grains (wheat, rice, barley, oats, rye and corn) are anemophiles. Up to 70% of the world's farmland is planted with grains yielding nearly 2 billion tonnes of food. Without the wind we would soon starve. Anemophiles produce huge amounts of pollen which allows cross-fertilisation up to 1200 km away in the case of pine and fir pollen. Spores and seeds also need to be dispersed and several travel by air. Fungal spores can circumnavigate the globe. The average edible mushroom can produce about 16 billion spores at the rate of 100 million an hour. Seeds have evolved a range of aerodynamic features resembling parachutes, propellers and wings. Dandelion (*Taraxacum officinale*) seeds with their fluffy umbrellas can ride the wind for up to 200 km. In the case of the Tumbleweed (*Salsola kali* or *Salsola tragus*) the entire plant is carried by the wind scattering its seeds along the way.

Another neglected atmospheric resource for communication is sound. Most people are totally unaware that sound travels *via* air molecules and waves in the atmosphere to our ears. Even in the ipod age of head phones the music has to travel to our inner ears *via* the local air molecules. The global music industry itself is worth at least $40 billion per year in the sales of music media.

It is very difficult to put a monetary value on everyday sounds and speech as used by everyone to communicate using mobile phones *etc*. Similarly the transmission of radio waves by the upper atmosphere (the ionosphere) is very important as it allows signals to travel around the globe. *Skywave* is the propagation of electromagnetic waves refracted back to the Earth's surface by the ionosphere.

6.9   Direct and Indirect Use of the Atmosphere for Energy and Power

A piece of air the size of a sugar lump contains around 25 billion billion molecules all constantly darting about faster than the speed of sound. Every molecule crashes into another 5 billion times a second and it is this incessant pinball barging that gives air its spring. It's why the billions of bouncing molecules inside a tyre can hold up a truck.[6]

Wind power is the most obvious direct form of atmospheric energy but wave power and hydroelectric power sources are indirectly reliant on the atmosphere. Solar power is controlled by the presence or absence of cloud and is therefore also dependent indirectly upon the atmosphere. Biofuels rely upon photosynthesis and oxygen for production.

In 2007 the wind industry installed close to 20 GW worldwide. This development was led by the US, China and Spain, and it brought global installed capacity to 94 GW. This is an increase of 31% compared with the 2006 market, and represents an overall increase in global installed capacity of about 27%. The top five countries in terms of installed capacity are Germany (22.3 GW), the US (16.8 GW), Spain (15.1 GW), India (7.8 GW) and China (5.9 GW). In terms of economic value, the global wind market in 2007 was worth about €25 billion, or US$37 billion in new generating equipment, and attracted €34 billion ($50.2 billion) in total investment. Europe remains the leading market for wind energy and new installations represented 43%t of the global total in 2007.[64]

Wind energy on-shore capacity is increasing rapidly as the technology matures, but siting, plus proximity to point of use, issues remain. MacKay[65] estimates that if the windiest 10% of Britain was covered in windmills (generating 2W m^{-2}) it would be possible to generate 20 kWh day^{-1} per person which is only half of the power used to drive an average fossil fuelled car 50 km per day. Off-shore wind power has more potential long term than on-shore. The UK government has committed to generating 33 GW of off-shore wind capacity. MacKay[65] estimates that this would require 10 000 '3 GW' (on average producing 1 GW each) windmills at a cost of about £33 billion over the next 10 years. Wind power could feasibly supply more than 20% of world energy requirements by 2050. This could require up to 2 million new windmills worldwide (50 times today's capacity), but could save up to 1 gigatonne of CO_2 emissions from coal-burning power stations.

Solar energy is used in a variety of ways: for heating water, photovoltaic panels for generating electricity, solar biomass for generating biofuels or burning directly in a power station, or for cooking. The potential is huge but often prohibitively expensive. To place photovoltaic panels in desert areas (cloud reduces output by up to 90%) would be the most effective. MacKay[65] estimates that:

One billion people in Europe and North Africa could be sustained by country-sized power facilities in deserts near the Mediterranean; and that half a billion in North America could be sustained by Arizona-sized facilities in the deserts of the USA and Mexico.

Waves are driven by the wind and there is a huge potential for the future but wave power is still in its infancy and still at the research stage. Hydropower is much more established and is dependent on rainfall. There are only a limited number of potential sites around the world that have not already been tapped. Worldwide, hydroelectricity supplied an estimated 715 000 MW of electricity in 2005. This was approximately 19% of the world's electricity and accounted for over 63% of electricity from renewable sources.

Air-source heat pumps are very efficient and rely on the air temperature difference between the inside and outside of buildings. A typical air-source unit today can deliver more than 4 kW of heating when using just 1 kW of electricity, with a coefficient of performance (COP) of 4 + . When it is run in reverse for air conditioning, it can deliver nearly 4 kW of cooling for 1 kW of electricity. The air can also be used to heat water for under-floor heating systems and radiators. The efficiency of air-source heat pumps is increasing and systems already on sale in Japan have a COP of 6.6.

We should replace all our fossil-fuel heaters with electric-powered heat pumps. We can reduce the energy required to 25% of today's levels. Heat pumps are future proof, allowing us to heat buildings efficiently with electricity from any source.[65]

Compressed air is used for a multitude of activities, including energy for transport and inflating tyres. Compressed air is a way of storing energy but obviously it needs energy to be compressed. The efficiency is similar to lead–acid batteries, but about five times less than lithium ion batteries:

Air can be compressed thousands of times and doesn't wear out . . . compressed air storage systems do have three advantages over batteries: longer life, cheaper construction and fewer nasty chemicals.[65]

6.10 The Extraction of Atmospheric Gases

There is a large global industry extracting atmospheric gases, such as oxygen, nitrogen, argon and other rare gases, to use as commodities for a huge range of industries. There is currently no charge or licence required for the removal/mining of gases from the air as raw materials – a blatant case of commodification of the atmosphere. Recently Carbon Capture and Storage (CCS) has been developed to try and directly reduce the amount of carbon dioxide in the atmosphere.

Table 5 Typical applications for gas products.

Markets	Applications	Customer Benefits	Products
Metal Fabrication	Welding & cutting	Productivity, Cost Reduction Product Quality	Argon
Primary Metals	Combustion in electric arc furnaces	Productivity, Energy Savings	Oxygen
	Refining of metal products	Product Quality	Argon
	Cast iron melting in rotary furnaces	Flexibility, Cost Reduction	Oxygen
Chemicals	Vent-gas recovery of volatile organics	Environmental	Nitrogen
	Vapor and liquid phase oxidation	Productivity, Cost Reduction	Oxygen
Petroleum Refining	Catalyst regeneration	Productivity	Oxygen
	Enhanced oil and gas recovery	Productivity	Nitrogen
	Gasoline reformulation	Environmental	Hydrogen
Food	Freezing and chilling	Productivity, Flexibility, Shelf Life, Flavor, Safety	Nitrogen, Carbon dioxide
	Packaging	Quality, Shelf Life	Nitrogen
Beverage	Carbonation	Quality	Carbon dioxide

The global market for gases is huge and companies that originally started by just extracting gases from the atmosphere (*e.g.* British Oxygen who have since merged with Linde) now manufacture many other gases (see Table 5). Air as the raw material is free and does not require the permits that a mine on land or sea would require.

6.11 Atmospheric Recreation and Climate Tourism

In 2008, international tourist arrivals reached 924 million, up 16 million over 2007, representing a growth of 2%, and although levels have fallen back in 2009 due to the global recession, they are picking up again according to the United Nations World Tourism Organisation.[81] Total expenditure was estimated in 2007 to be $856 billion, which represents an expenditure of nearly $1000 per tourist per year. Not all tourism is weather/climate related but certainly a large proportion of summer and winter tourism (skiing) relates to international travel and the seeking of a warmer/dryer climate.

All outdoor and indoor sport is affected by the atmosphere, weather and climate.

There is still much confusion as to how the atmosphere (*e.g.* air density) controls the enjoyment of a sport. For example, sports such as squash and badminton are normally played indoors in artificial 'ecoclimates' created by sports centres to avoid wind and rain. However, air temperature and humidity are still of vital importance.

Thornes[68] defines three types of sport that are weather dependent: Specialised Weather Sports (*e.g,* sailing, gliding, ballooning and skiing), Interference Weather Sports (*e.g.* lawn tennis, football, rugby and hockey) and Weather Advantage Sports (*e.g.* golf, baseball and cricket).

The atmosphere is an integral part of the game of cricket.[69] For example, snow stopped play at Buxton in June 1975 during the game between Derbyshire and Lancashire. Lancashire scored 477–5 in 100 overs on the Saturday. It then snowed on the Sunday and play was abandoned for the day on Monday. However, the snow melted quickly and play was possible on the Tuesday. In the days of uncovered wickets, Derbyshire were bowled out for 42 and 87 and Lancashire won by an innings and 348 runs! This is one of the biggest ever victory margins recorded in English County Cricket history.

6.12   Aesthetic, Spiritual and Sensual Properties

The atmosphere is a canvas for heart-stopping displays of beauty.[31]

For most of the time the atmosphere seems to be odourless but we need the atmosphere to transfer the smell of a rose or a dead rat to our senses. Smells are composed of small volatile molecules that move by diffusion – random walks – to be breathed in by our receptive noses. Humans have about 5 million receptor cells (dogs have about 200 million) waiting to communicate with our memory banks to recognise a smell. Hundreds of thousand molecules exist that have different smells and different cultures can perceive the pleasant and the awful smell differently. The global size of the perfume market was estimated to be $18 billion in 2006.[70] There is some evidence that pleasant smells make us feel good, and hence the concept of aromatherapy as an alternative medicine. Taste is also based on smell as about 75% of the flavour of food is linked to the smell. Hence smell is a very important part of our quality of life. In the animal kingdom social behaviour is based on smell:

Based on odour, animals zero in on food, avoid predators, recognize the boundaries of their home territory, identify friends and family and likely members of the opposite sex as prospective mates. Odours used to convey informational signals from individual to individual are collectively known as semiochemicals. Corn, beets and cotton, for example, infested with leaf-munching larvae of the beet armyworm moth, put up a fight, throwing out insidious indoles and terpenes – chemicals that attract parasitic wasps. The wasps lay their eggs in armyworm caterpillars, with fatal results; the young wasp larvae eat their host as they hatch.[4]

Pheromones are a subset of semiochemicals that pass information within an individual species. Social insects like ants, bees, wasps and termites communicate between themselves *via* a large range of pheromone signals. However, pheromones secreted by the female gypsy moth are simulated by the bola spider, luring male gypsy moths to be eaten.

The sky is a daily source of ephemeral environmental art that is visible to all around the world. The visualisation of atmosphere, weather, climate and climate change in art works is a fascinating component of representational Environmental Art and Cultural Climatology.[71,72] Also non-representational performative works of artists, such as James Turrell's *Skyscapes,* focus upon the sky and atmosphere and the *Skyscapes* have created a wonderful metaphor, and lever for action, in restoring the 'hole in the ozone layer'. The atmosphere has been an inspiration for poets and writers as well as artists. Jacobus[73] in discussing the poetry of Clare and the paintings of Constable declares:

Clouds are the realm of the visible invisible, both what we can and what we can't see; their representation involves the double relation of the work of perception and the work of art, along with our complex, feeling, yet pre-determined relation to both.

Rehdanz and Maddison[74] suggest that different people have preferences for different types of weather and climate and that happiness and well-being are directly related to the state of the atmosphere. A number of recent studies have emphasised the links between climate and human development.[75–79]

7   Conclusions

It is inevitable that the atmosphere should finally be recognised as the Earth's most important and valuable natural resource in this time of climatic change. It is vital that the atmosphere is effectively managed to sustain its composition and delicate balance.[5]

The sustainable management of the atmosphere needs to be tackled from a global perspective. Some intervention policies to protect the global atmospheric commons have already been effectively implemented, on a local, regional and global scale. Local air quality legislation in the USA and Europe has systematically reduced key air pollutants since the 1990s, principally to protect public health in cities. The Gothenburg protocol, E.U. Emissions Ceilings Directive, and the Convention on Long-Range Transport of Air Pollution (CLRTAP) are only some of the regional policies successfully put in place to prevent regional air quality effects, such as acid rain and ground-level ozone generation. The Montreal Protocol has shown that decisive and sustainable management of a global atmospheric problem is also possible, within relatively short timeframes. The Kyoto Protocol however has been less successful at producing a sustainable solution to enhanced global warming. This chapter aims to catalyse action for more decisive sustainable management of the atmosphere in the future.

The valuation of Atmospheric Services at between 100 and 1000 GWP is undoubtedly an underestimate and further research is required to give more realistic valuations for each of the twelve individual atmospheric services. In order to effectively manage the atmosphere we need to recognise which services are at risk and this also requires further research. It is clear, however, that the atmosphere is our most precious natural resource and that it needs careful protection from exploitation and commodification. The atmosphere should be declared as a global commons that belongs to everyone and a Law of the Atmosphere needs to be ratified by the United Nations as soon as possible. At a time of enhanced global warming and the real prospect of geo-engineering of our climate, we need swift action to communicate the need to preserve and protect the atmosphere for our future well-being on the planet.[82]

Climate change and increases in severe weather, heatwaves, floods, droughts, rising sea levels, melting ice, and consequential elevated mortality and threats to biodiversity are seen by policy makers as a huge problem. These impacts can only be managed effectively if we understand the complete system and value of atmospheric services.

Acknowledgements

I wish to thank the following colleagues who have attended workshops and commented on and suggested improvements to this chapter (any mistakes are my own however): Bill Bloss, Stefan Buzar, Xiaoming Cai, Lee Chapman, Julian Clark, Suraje Dessai, Sen Du, Ian Fairchild, Dan van de Horst, Michaela Kendall, Chris Kidd and Sam Randalls.

References

1. J. Ruskin, *The Queen of the Air,* George Allen, 1887, p. 239.
2. J. Ausubel, in *Climatic Constraints and Human Activities*, ed. J. Ausubel and A. K. Biswas, IIASA Proceedings Series, Pergamon Press, Oxford, 1980, **vol. 10**, pp. 13–59.
3. P. Barnes, *Who Owns the Sky?,* Island Press, London, 2001, p. 172.
4. R. Rupp, *Four Elements: Water, Air, Fire and Earth,* Profile Books, London, 2005, p. 373.
5. J. E. Thornes and S. Randalls, *Geograf. Annal.*, 2007, **89A**, 273–285.
6. G. Walker, *An Ocean of Air: A Natural History of the Atmosphere,* Bloomsbury Press, London, 2007, p. 321.
7. R. Costanza, *et al., Nature*, 1997, **387**, 253–260.
8. Munich Re, *Topics Geo*, Munich, 2009, p. 45.
9. N. Stern, *Blueprint for a Safer Planet,* The Bodley Head, London, 2009, p. 246.
10. J. Lovelock, *Gaia, A New Look at Life on Earth,* Oxford University Press, Oxford, 1979.
11. J. Lovelock, *The Revenge of Gaia,* Penguin, London, 2006, p. 222.

12. World Bank, *Weather and Climate Services in Europe and Central Asia*, Washington D.C., 2008, Working Paper No. 151, p. 80.
13. X. Xu, *Report on Surveying and Evaluating Benefits of China's Meteorological Service,* China Meteorological Administration, Bejiing, China, 2007.
14. M. Visbeck, *Nature Geosci.*, 2008, **1**, 2–3.
15. C. Hindmarch, J. Harris and J. Morris, *Biologist*, 2006, **53**, 135–142.
16. R. Haines-Young and M. Potschin, *England's Terrestrial Ecosystem Services and the Rationale for an Ecosystem Approach: Overview Report to DEFRA (DEFRA Project Code NR0107)*, Defra, UK, 2008.
17. R. K. Craig, *The Atmosphere, the Oceans, Climate and Ecosystem Services*, 2008. http://ssrn.com/abstract = 1114354 (accessed on 12/03/10).
18. J. Boyd and S. Banzhaf, *Ecol. Econ.*, 2006, **63**, 616–626.
19. K. J. Wallace, *Biol. Conserv.*, 2007, **139**, 235–246.
20. R. Costanza, *Biol. Conserv.*, 2008, **141**, 350–352.
21. Millennium Ecosystem Assessment, *Ecosystems and Human Well-Being – Our Human Planet*, Island Press, Washington DC, 2005, p. 137.
22. W. J. Maunder, *The Value of Weather*, University Paperback, London, 1970.
23. J. McNeill, *Something New Under the Sun*, Penguin Books, London, 2000, p. 421.
24. J. E. Thornes and G. R. McGregor, in *Contemporary Meanings in Physical Geography*, ed. S. Trudgill and A. Roy, Springer, London, 2003, ch. 8, pp. 173–197.
25. Air Resources Board, *How Much Air Do We Breathe?*, 1994, Research Note 94-11.
26. M. Hulme, *Why We Disagree about Climate Change*, Cambridge University Press, Cambridge, 2009, p. 392.
27. J. E. Thornes and L. Chapman, *Geogr. Compass*, 2008, **2**, 1012–1026.
28. *J. Atmos. Environ.* published two papers on 'New Directions' for the new millennium: A. Najam, *Atmos. Environ.*, 2000, **34**, 4047–4049 and J. Vogler, *Atmos. Environ.*, 2001, **35**, 2427–2428.
29. A. Balkin, 2007. http://www.publicsmog.org (accessed on 7/4/09).
30. W. H. Schauer, *J. Int. Affairs*, 1977, **31**(1), 67–80.
31. S. J. Buck, *The Global Commons*, Earthscan, London, 1998, p. 225.
32. O. Tickell, *Kyoto2*, Zed Books, London, 2008, pp. 293.
33. G. Hardin, *Science*, 1968, **162**(3859), 1243–1248.
34. N. E. Bockstael, A. M. Freeman, R. J. Kopp, P. Portney and V. K. Smith, *Environ. Sci. Technol.*, 2000, **32**, 1384–1389.
35. G. C. Daily, T. Soderqvist, S. Aniyar, K. Arrow, P. Dasgupta, P. R. Ehrlich, C. Folke, A.-M. Jansson, B.-O. Jansson, N. Kautsky, S. Levin, J. Lubchenco, K.-G. Mäler, D. Simpson, D. Starrett, D. Tilman and B. Walker, *Science*, 2000, **289**, 395–396.
36. G. Atkinson and S. Mourato, *Annu. Rev. Environ. Resour.*, 2008, **33**, 317–344.
37. CIA World Fact Book, Central Intelligence Agency Economy section, 2008.

38. D. A. Starrett, *Valuing Ecosystem Services*, 2003. http://www.undp.org.cu/eventos/instruverdes/envvaliationDStarret.pdf (accessed 12/03/10).
39. N. Stern, *Stern Review on The Economics of Climate Change*, Cambridge University Press, 2006. http://www.cambridge.org/9780521700801.
40. Defra, *An Introductory Guide to Valuing Ecosystem Services*, 2007, p. 65.
41. R. A. Kramer, in *The Sage Handbook of Environment and Society*, ed. J. Pretty, A Ball, T. Benton, J. Guivant, D.R. Lee, D. Orr, M. Pfeffer and Prof. H. Ward, Sage, London, 2007, ch. 11, pp. 172–180.
42. N. P. Hyslop, *Atmos. Environ.*, 2009, **43**, 182–195.
43. H. Welsch, *Ecol. Econ.*, 2006, **58**, 801–813.
44. P. W. Boyd, *Nature Geosci.*, 2008, **1**, 722–724.
45. R. B. Cathcart and M. M. Æirkoviæ, in *Macro-Engineering*, ed. V. Badescu, R. B. Cathcart and R. D. Schuiling, Springer, The Netherlands, 2006, ch. 9.
46. Royal Society, *Geoengineering the Climate*, 2009, RS1636 London.
47. A. Robock, A. Marquardt, B. Kravitz and G. Stenchikov, *Geophys. Res. Lett.*, 2009, 36, L19703, doi:10.1029/2009GL039209.
48. M. Merleau-Ponty, *Phenomenology of Perception* (transl. by C. Smith), Routledge and Kegan Paul, London, 1962.
49. N. Lane, *Oxygen: The Molecule that made the World*, Oxford University Press, Oxford, 2002, p. 374.
50. J. M. Wallace and P. V. Hobbs, *Atmospheric Science*, Elsevier, Amsterdam, 2006.
51. S. Hoffman, *Planet Water*, John Wiley & Sons, 2009, p. 44.
52. R. G. Prinn, *Annu. Rev. Environ. Resour.*, 2003, **28**, 29–57.
53. World Health Organisation, *Air Quality Guidelines – Global Update 2005*, 2006.
54. D. W. Dockery, C. A. Pope 3rd, X. Xu, J. D. Spengler, J. H. Ware, M. E. Fay, B. G. Ferris Jr and F. E. Speizer, *New Engl. J. Med.*, 1993, **329**(24), 1753–1759.
55. P. Wilkinson, K. R. Smith, M. Joffe and A. Haines, *The Lancet*, 2007, **370**, 965–978.
56. W. J. Gauderman, E. Avol, F. Gilliland, H. Vora, D. Thomas, K. Berhane, R. McConnell, N. Kuenzli, F. Lurmann, E. Rappaport, H. Margolis, D. Bates and J. Peters, *New Engl. J. Med.*, 2004, **351**, 1057–1067.
57. A. J. Cohen, H. R. Anderson, B. Ostra, K. D. Pandey, M. Krzyzanowski, N. Künzli, K. Gutschmidt, A. Pope, R. Romieu, J. M. Samet and K. R. Smith, *J. Toxicol. Environ. Health*, 2005, **68**, 1–7.
58. D. Reay, *Planet Earth*, Summer 2009, 14–17.
59. FMB Group Ltd, *FMB Weekly Nitrogen Report*, March 2009.
60. UK Department of Transport (UKDFT) *Highways Economics*, Note No. 1, 2007.
61. S. Gössling and P. Upham, *Climate Change and Aviation*, London, Earthscan, 2009, p. 386.
62. International Air Transport Association, *Building a Greener Future*, 2nd edn, 2009. http://www.iata.org/nr/rdonlyres/c5840acd-71ac-4faa-8fee-00b21e9961b3/0/building_greener_future_oct08.pdf

63. ATW Eco-Aviation, http://www.atwonline.com/channels/eco/article.html? articleID = 2721
64. BWEA (British Wind Energy Association) website (bwea.com).
65. D. J. C. MacKay, *Sustainable Energy – Without the Hot Air,* UIT Press, Cambridge, 2009, p. 366.
66. S. T. Coleridge, *The Rime of the Ancient Mariner,* 1798, in *The poems of Samuel Taylor Coleridge,* ed. E. H. Coleridge, Oxford University Press, London, 1917.
67. G. Walker and D. King, *The Hot Topic,* Bloomsbury, London, 2008, p. 309.
68. J. E. Thornes, *Weather,* 1977, **32**, 258–268.
69. J. E. Thornes, *Area,* 1976, **8**(4), 105–112.
70. *Wall Street Journal,* "Why the Perfume Industry is Beginning to Stink", December 24th 2007.
71. J. E. Thornes, *Geoforum,* 2008, **39**, 570–580.
72. J. E. Thornes, *Annu. Rev. Environ. Resour.,* 2008, **33**, 391–411.
73. M. Jacobus, *Gramma,* 2006, **14**, 219–247.
74. K. Rehdanz and D. Maddison, *Ecol. Econ.,* 2005, **52**, 111–125.
75. E. Durschmied, *The Weather Factor,* Coronet, London, 2000.
76. V. Jankoviæ, *Reading the Skies,* University of Manchester Press, Manchester, 2000.
77. R. Hamblyn, *The Invention of Clouds,* Picador, Oxford, 2001.
78. L. Boia, *The Weather in the Imagination,* Reaktion, London, 2005.
79. R. Costanza, L. Graumlich, W. Steffen, C. Crumley, J. Dearing, K. Hibbard, R. Leemans, C. Redman and D. Schimel, *Ambio,* 2007, **36**(7), 522–527.
80. World Health Organisation (WHO) http://www.who.int/int/mediacentre/ factsheets/fs313/en/index.html (accessed 24/03/10).
81. United Nations World Tourism Organisation (UNWTO), http:// unwto.org/facts/eng/barometer/UNWTO_Barom09_3_excerpt.pdf (accessed 24/03/10).
82. J. E. Thomas, W. Bloss, S. Bouzarovski, X. Cai, L. Chapman, J. Clarke, S. Dessai, S. Du, D. van der Horst, M. Kendall, C. Kidd and S. Randalls, *Met. Apps.,* 2010, (in press).

Natural Capital and Ecosystem Services: The Ecological Foundation of Human Society

ERIK GÓMEZ-BAGGETHUN AND RUDOLF DE GROOT

ABSTRACT

Ecosystems provide both the energy and materials needed for the production of economic goods and services and act as a sink of wastes
generated by the economic metabolism. Other nature's services benefits
are obtained directly from nature, often without passing through transformation processes or the mediation of markets, as in the case of clean
air, erosion control, aesthetic benefits, or climate regulation. Economic
health in the long term thus depends on the maintenance of the integrity
and resilience of the natural ecosystems in which it is embedded. The fact
that standard economic theory neglects this aspect has been identified as a
main cause of the current environmental problems and ecological crises.
Approaches such as ecological and environmental economics attempt to
deal with these shortcomings of standard economics through the development of concepts and accounting methods that better reflect the role of
nature in the economy and the ecological costs derived from economic
growth. Concepts such as natural capital, ecosystem functions and ecosystem services are playing a key role as tools to communicate societal
dependence on natural ecosystems and in the articulation of a new form
of understanding economics. This paper gives a brief overview of the key
concepts for understanding the links between ecosystems and human
well-being, and discusses a range of valuation and accounting methods as
possible ways to measure the quantities and importance of natural capital
and ecosystem services.

Issues in Environmental Science and Technology, 30
Ecosystem Services
Edited by R.E. Hester and R.M. Harrison
© Royal Society of Chemistry 2010
Published by the Royal Society of Chemistry, www.rsc.org

1   Introduction

In his 1926 book, *Wealth, Virtual Wealth and Debt*, the English biochemist Frederick Soddy (awarded the Nobel prize for Chemistry in 1921 for his formulation of the theory of isotopes) criticized the focus of economics on monetary flows, arguing that 'real' wealth is primarily derived from the use of energy to transform materials into physical goods and services.[1] His observation represents a pioneer formulation of the modern ecological criticism of the focus of economics on the monetary system at the expense of the analysis of the interactions with the biophysical system in which it is embedded.[2,3] Indeed, Soddy's scepticism of the way financial capital develops a life of its own and becomes increasingly decoupled from the underlying physical capital, gains renewed interest in the context of the present economic crisis.[4]

Since the last two to three decades, Soddy's intuition has been developed and formalized within the field of ecological economics. This approach maintains that all the goods and services produced in the economic system ultimately depend on inputs of energy and materials from ecosystems. On the other hand, this approach maintains that natural ecosystems constitute the sinks for all wastes resulting from economic activity.[5,6] In addition to providing these core source and sink functions, ecosystems provide a wide variety of benefits to people which often are obtained directly from nature without passing through transformation processes or the mediation of markets, as in the case of clean air, water purification, climate regulation, or erosion control.[7,8] The fact that standard economic theory neglects this aspect has been identified as a main cause of current environmental degradation.[9] A key factor that has been pointed out in this respect is the limited capacity of current economic accounting systems to reflect physical costs derived from economic growth in terms of, for example, pollution or physical depletion. Scientific approaches, such as environmental economics and ecological economics, have put considerable efforts in reconnecting economic systems with the underlying ecological systems. Environmental economics aims to monetise environmental externalities to incorporate them into current economic accounting systems. Ecological economists question the foundations and axioms on which such accounting systems rely, and attempt to develop new conceptual and analytical frameworks capable of better incorporating the role of ecosystem functions and services into economic theory and practice.

Although criticism of conventional economics is at least 120 years old,[10] economics has so far remained highly reluctant to revise its theoretical foundations to better incorporate the economic importance of ecosystems within its analytical and accounting framework,[11] Only recently, the need to integrate ecological and economic accounts began to find its way into the international policy agenda,[12] An important example is the ongoing development of a System of integrated Environmental and Economic Accounts (SEEA) under the auspices of the United Nations, which eventually will provide guidelines for the amendment of the existing systems of national economic accounts.[13]

Much of the present policy interest in the revision of economic accounting systems is the result of efforts during the last three decades devoted to the development of concepts, like natural capital and ecosystem services, that better reflect the role of nature in the economy. First used in the 1970s as metaphors to communicate societal dependence on natural ecosystems, in the last decade the literature around natural capital and ecosystem services has grown exponentially,[14] becoming a field of research on its own sometimes referred to as 'ecosystem-service research' or 'ecosystem-service science'.[15,16]

2 Societal Dependence on Ecosystems in Different Socio-Economic Contexts

Societal dependence on ecological life-support systems is evident in subsistence economies where communities depend directly on ecosystems for food and other products needed for their livelihood. Nevertheless, countries with developed market economies also depend on ecosystem services in many ways. However, this dependence is often overseen for a series of reasons. First, in the so-called developed countries, economic activities that rely intensely on natural resources are becoming increasingly marginal due to the current offshoring process through which extractive and industrial activities are being relocated in less-developed countries, with lower labour costs and softer environmental legislation.[17] Second, in developed-country settings, most ecosystem services are not directly obtained or enjoyed from nature as occurs in subsistence economies, but are 'embedded' in market products (*e.g.* imported food). Furthermore, such services are often obtained from 'anonym' ecosystems in distant countries after going through multiple stages of the transformation and distribution chains. In this manner, the ecological contribution to the end-product becomes masked by an increasingly de-localized economic process, alienating the consumer from the links between the source ecosystems and the final goods and services that are consumed or enjoyed.

This economic context has fostered the idea within developed countries that, by means of technology and markets, socio-economic systems can progressively 'dematerialize',[18] thus becoming increasingly decoupled from natural ecosystems. The theory of dematerialization of the economy is often based on the so called Kuznets Curve hypothesis,[19] which maintains that once a certain level of national wealth is surpassed, economic growth becomes less intensive in pollution and decreasingly dependent on natural capital and the services it provides.[20] However, the validity of this hypothesis has been widely questioned[18,21] when not openly refuted by studies that suggest that economic growth remains highly coupled to energy and material input, even in developed country settings.[17,22]

Factors such as modern technology, the expansion of the tertiary sector, offshoring, and the presence of markets as mediators for the enjoyment of ecosystem services, have led to the idea that socio-economic systems can be decoupled from the ecological systems in which they are embedded. However, such ecological–economic decoupling can only occur at the local scale. Ultimately, every good and service produced by the economic system depends on transformations of energy and materials that, at least at the present stage of technological development, solely nature can provide.[1,5,23] Moreover, the extent to which technological development can help overcome the problem of ultimate physical scarcity remains controversial.[24] Whereas standard economic theory maintains that technology will eventually allow for the progressive substitutability of scarce natural resources,[25,26] ecological economics maintains that this assumption is at odds with the laws of thermodynamics.[2,23]

The fact that developed countries can satisfy ever increasing consumption demands and at the same time decrease the exploitation intensity within their territories is not due to the economy's dematerialization, but rather because the present international free-trade system allows wealthy consumers from these countries to obtain ecosystem services from all around the world through the globalised market.[27] Gross Domestic Product growth in developed countries is thus only possible because of being supported by ecosystem source functions (*e.g.* oil) and sink functions (*e.g.* atmosphere) provided by natural capital assets located mainly beyond their borders.

As biodiversity loss continues and the feeling expands that ethics-based arguments for conservation are reaching their limits, emphasising the dependence of human well-being on ecosystem services provides an important additional rational for nature conservation.[15,28] The Millennium Ecosystem Assessment (MA) project has openly adopted this approach.[9] Conservation and sustainable use of natural ecosystems is thus no longer presented simply as an ethical duty towards future generations, nor as a *luxury goal* that only rich countries can afford to address, as reflected by the post-materialist theories with great influence in the 1980s and 1990s.[29] Now, ecosystems are also acknowledged as the basis of the economic and social development on which human well-being relies.

3   Understanding the Links between Ecosystems and Human Well-Being

If analyzed through an anthropocentric lens, ecosystems are both essential life-support systems and economic-support systems. From this perspective, ecosystems can be seen as a form of (natural) capital. The notion of Natural Capital has a clear precursor in the land production factor which classical economists used to include in their production functions,[6,30] but its first explicit mentioning dates to the 1970s.[31] It was not until the 1990s that the natural capital concept was developed in the literature,[5,32,33] after which it has become widely used in the fields of environmental and ecological economics. Costanza and Daly[5] draw on an analogy of the economic definition of capital to coin natural capital as a '*stock* capable to provide a sustainable flow or *natural income* (ecosystem services)', a definition that, with minor variations, has persisted in the literature until today.

Drawing on these metaphors or ecological–economic hybrid concepts, the MA developed a comprehensive framework in which ecosystems are portrayed as natural capital stocks providing flows of ecosystem services.[9] The MA framework represented a great progress for the understanding of the links between ecosystems and human well-being. However, the MA framework needs to be further developed to make it operational for specific purposes and applicable in 'real life'. For instance, scholars concerned with the incorporation of ecosystem services in economic accounting systems highlighted that the causal chain between ecosystem properties (including biodiversity) and the final output of ecosystem services is generally left as a black box in the ecosystem-service literature.[34] Subsequently, they call for the need to disentangle the sequence of causal links from ecosystems structure and processes to the final benefits that are enjoyed, consumed or used by humans.

The debate on an operational classification of ecosystem functions, intermediate services and final services is still ongoing.[34-36] Controversial aspects in this debate include whether we should aim for a generic, universally accepted classification of ecosystem services *versus* maintaining several classifications for different purposes;[37] whether human inputs to realise nature's benefits should be subtracted when quantifying ecosystem services; and what distinction should be made between 'functions', 'services' and 'benefits' (and 'values'). In relation to the latter point, the growing consensus now seems to be that there is a distinction which can be seen as a 'cascade' going from the ecosystem properties *via* functions, to services which provide benefits and values.[36,38,39]

Drawing on our own previous work[8,40,41] and on recent developments on this topic,[36,39,42] we hereby provide a framework to understand the links between ecosystem structures and processes, and the final output of ecosystem services. Figure 1 provides a framework showing a gradient from ecosystem structure and functioning to the final benefits relating directly to human well-being.

More concretely, our framework distinguishes four different levels of analysis (see Table 1). The first level includes the totality of ecological components (structure) and processes (functioning) operating within an ecosystem. The second involves Ecosystem Functions, which is a subset of ecological processes and components that are directly involved in the underpinning of ecosystem services. The third are the Ecosystem Services, which are actively used, enjoyed, or consumed at a given point in space and time. Finally, the last level of analysis relates to the *benefits* from ecosystem services or the relative impact of ecosystem services on well-being as perceived by humans. Each level is explained in more detail in the sections below.

3.1 Ecosystem Structure and Functioning

The first level of the ecosystem's human well-being framework includes the whole set of ecological components (*e.g.* matter, energy, and species) and core ecological processes (*e.g.* nutrient and energy cycling) operating within the ecological system. Each of these processes and components (conceptualized in the MA framework as 'supporting services') constitute a more or less relevant role in maintaining the basic functioning of the ecosystem.

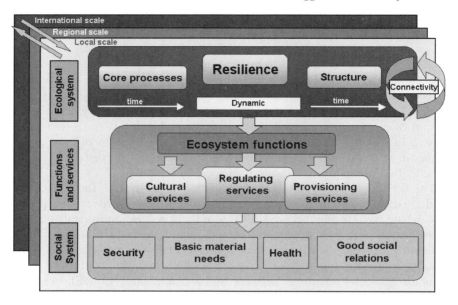

Figure 1   Natural capital and human well-being. Ecosystem functions enable the provision of diversified flows of ecosystem services, which in turn influences the different components of human well-being.

Table 1   Levels of analysis to understand the links between ecosystems and human well-being.

1	Ecosystem structure and processes ('functioning')	Totality of components and processes operating in the ecosystem
2	Ecosystem functions	Subset of components and processes that are involved in the generation of ecosystem services
3	Ecosystem services	Subset of ecosystem functions that are actively or passively used, consumed or enjoyed by humans consciously or unconsciously
4	Ecosystem benefits	Value of ecosystem services as perceived by humans

A key property emerging from the structure and function interplay is the resilience of the ecosystem. Resilience relates to the capacity of the ecological system to withstand shocks while maintaining the essential structure and functioning.[43] It thus reflects the ecosystem's capacity to self-organize in order to adapt to disturbance and change.[44] Resilience is of critical relevance in the context of ecosystem services, as it provides the ecological system with the capacity to maintain ecosystem service flows within tolerable bounds throughout time in the face of variability and change.[45] If human pressure on ecosystems exceeds certain thresholds, ecosystems may flip to alternative stable states, often with far less capacity to provide ecosystem services.[46,47]

3.2   Ecosystem Functions

Once the structure and functioning of an ecosystem has been characterized, the next step in assessing the links between ecosystem and human well-being is to translate ecological complexity (total set of ecological processes and components) into a limited set of ecosystem functions, which in turn provide a range of goods and services.[8,48]

In the ecological–economic context addressed here – not so in ecology – ecosystem functions can be defined as those ecological processes and components with the *capacity* to generate benefits that are enjoyed directly or indirectly by humans.[8,49,50] Ecosystem functions thus represent the subset of the total ecological components and processes that are directly involved in underpinning ecosystem services, and thus refer to the ecological capacity to support economic activities. In this sense, ecosystem functions represent the key link between ecology and economy,[49] and provide an essential conceptual tool for the further development of a natural capital theory with sound ecological basis.

3.3   Ecosystem Services

Ecosystem services refer to those ecosystem functions that are actively used, enjoyed or consumed by humans, which can range from material goods (such as water, raw materials and medicinal plants) to various non-marketed, and therefore 'free', services (such as climate regulation, waste assimilation, water purification, carbon sequestration, erosion control, flood buffering, *etc.*).[7–9] What economics has traditionally referred to as goods and services has been reconceptualised within sustainability sciences in the broader concept of ecosystem services,[9,50,51] embedding all ecosystem goods and services, whether marketed or not, that directly or indirectly contribute to human well-being.

Thus, the *potential* benefits represented by ecosystem functions become *actual* benefits once they are demanded, used or enjoyed by people. It is then that functions acquire the form (now in a purely anthropocentric approach) of ecosystem services. Hence, it should be noted that the existence of ecosystem services is conditioned by the presence of beneficiaries of those services.[36,40] An uninhabited and unexploited forest, for example, will certainly have the potential for providing wood, but this would become a service only once someone went to the forest to cut the wood. In contrast, the carbon sequestration function provides a service even if the forest is uninhabited or unexploited because the global community benefits from its contribution to global climate regulation.

3.4   Ecosystem Benefits and Human Well-Being

The last level of analysis deals with the output from ecosystem services that link directly to human well-being. The measurement of the contribution of ecosystem services to human well-being is generally done through valuation processes. As it will be explained later, value elicitation exercises can use different valuation languages.[52] Whereas the valuation domain that has attracted most policy interest is the monetary one, there is a growing consensus on the need to approach valuation processes from alternative valuation domains, *e.g.* economic, socio-cultural and ecological, in order to capture the richness of values at stake in decision making processes.[8,48]

In order to avoid double-counting problems, economic valuation should be focused on the final contribution of ecosystems to human well-being, and in this respect several recent contributions have emphasised the need to distinguish between ecosystem services and ecosystem benefits.[36,39,42] This distinction allows the valuation process to be focused on the benefits of the service as perceived by humans in terms of their needs, rather than on the ecosystem service itself. This has several advantages. First, because, whereas most basic human needs are relatively stable, the value humans allocate to the services are more volatile and contingent on aspects like fashion, advertisement, market fluctuations, and culturally constructed 'wants'. Second, humans ultimately depend on ecosystem services irrespective of whether they perceive it or not, and whether they value ecosystem services or not, *e.g.* as expressed by their willingness to pay for their conservation.[53] Moreover, the value humans allocate to ecosystem services is contingent on the level of understanding of the way ecosystems are linked to well-being, and thus by the knowledge and characteristics of the people allocating the values.[54] As it has been argued earlier in this paper, in developed-country settings this link is largely obscured by the spatial mismatch between the production and the consumption of services and by alienation from nature as human populations are increasingly concentrated in cities.

Ecosystem benefits thus relate to the value of ecosystem services, which can be made explicit using particular measurement tools. Valuation approaches can be broadly divided between those approaches based on human preferences ('subjective' value approaches) and approaches based on physical costs ('objective' value approaches).[10,40,55] The bulk of the environmental accounting and valuation literature measures the benefits from ecosystem services using a 'human preferences-based approach', *i.e.* assuming that values can be derived from human subjective perception. Such subjective values are often elicited as weighted by their virtual prices or marginal willingness to pay. The second perspective to approach value aims to derive values from intrinsic characteristics of objects that can be measured in biophysical terms (embodied labour time, embodied energy or surface requirement for the production of the valuation subject). Here we will refer to this approach as 'physical costs-based approach'. In the following section, we review the main available methods within each of these two perspectives to approach ecosystem values.

4 Accounting and Valuation of Natural Capital and Ecosystem Services

Value theory has been referred to as 'the philosopher's stone of economics'.[56] Indeed, it is difficult to agree on a philosophical basis to value. In the context of ecosystem services, valuation can be viewed as a process of assigning 'weights' in decision-making processes. In their seminal paper on global natural capital valuation, Costanza and colleagues[50] suggested that the almost systematic undervaluation of the ecological dimension in decision making could be explained to a large extent by the fact that the services provided by natural capital are not adequately quantified in terms comparable with economic services and manufactured capital. Since this milestone publication, many academic efforts within sustainability sciences have worked on the development of accounting and valuation methods with the aim of highlighting the economic role of ecosystem services whose value was systematically undervalued or ignored by both markets and in decision making.

Nevertheless, a broad consensus on value theory has not been achieved yet. Its interpretations and diverse formulations rely on diverging epistemologies and methodological frameworks. Herewith, we present two main approaches of accounting and valuation within socio-ecological sciences, which we consider to be complementary and not exclusive (see Figure 2).

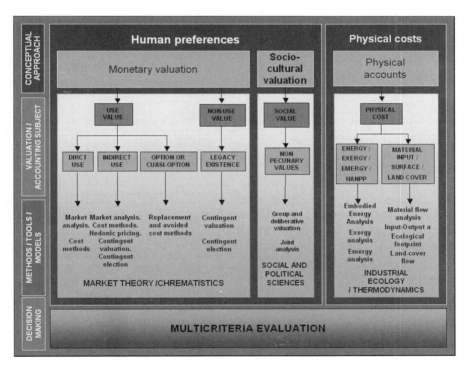

Figure 2 Different approaches for natural capital accounting and valuation. Value is multidimensional, and its estimation can be tackled from different perspectives. Multi-criteria analysis allows for the integration of different and incommensurable value types within decision making.

4.1   Approaches Based on Human Preferences

The first approach we refer to is the one based on human preferences. This approach aims to capture the intensity of people's preferences for small or marginal changes in the quantity or quality of these services. This conception of value is thus anthropocentric and instrumental in nature. Although preference-based approaches embrace methods that allow expressing individual (or group) preferences in non-monetary terms, *e.g.* by ranking among alternatives, pre-ference-based approaches usually take markets (prices) as reference for the allocation of values.

4.1.1   Market Theory Based Valuation

The neoclassical position is to view environmental problems and ecosystem services with no price as 'market failures'. Pure neoclassical economics restricts its accounting to priced goods and services, which entails considering just a limited subset of ecosystem services (those which are directly useful, valuable, appropriable and exchangeable). As price formation is conditioned to the existence of supply and demand relations, every positive or negative effect in human well-being lacking a market becomes invisible to conventional economic accounts. The public good nature of most ecosystem services implies that their economic value is often not adequately reflected in decision-making processes that are based on market transactions (*e.g.* cost–benefit analysis). As a consequence, it is argued, priceless ecosystem services are likely to be overutilized and tend to be depleted.

Valuation attempts to fill this void in the decision-making process. By assigning ecosystem services an economic value, it is possible to assess from an economic point of view whether a particular change in an ecosystem will improve or not (according to the neoclassical standards) the well-being of the community at which it is targeted. From this perspective valuation can be seen as a tool for trade-off analysis.[48]

The valuation of *invisible* costs and benefits, often referred to as 'external-ities', for their incorporation in economic accounting and decision making constitutes the cornerstone of environmental economics. With this aim, value types not captured by conventional markets are proposed (*e.g.* indirect use value, non-use value). Economists generally stick to taxonomy of ecosystem values which add up to the so-called Total Economic Value (TEV).[57] A range of methods has been developed in order to elicit ecosystem values. If available, information to elicit such values is obtained from market transactions (direct market valuation), but in their absence, information is derived from parallel markets (revealed preferences method) or hypothetical markets, *i.e.* markets created through surveys (stated preferences method). Some valuation methods are more appropriate than others for valuing particular ecosystem services and for the elicitation of specific value components.[58]

Environmental economics expands thus the analytical framework of pure neoclassical economics but without transcending the boundaries of chrema-tistics, that is, the field of monetary valuation.

4.1.2 Socio-Cultural Perception/Deliberative Based Valuation

Social perceptions and values play a fundamental role in the valuation people do in any decision-making situation, including those involving natural capital and its services. Aspects such as education, cultural identity and diversity, freedom, altruism and spiritual values have been pointed as human preferences shaping factors.[59–61] These approaches are not conditioned to monetisation in order to make options comparable, as decision making can be oriented through the ranking of preferences expressed by stakeholders after individual or group deliberation.

4.2 Approaches Based on Physical Costs

Biophysical valuation methods use a 'cost of production perspective' that aims to derive values from the physical costs (*e.g.* in terms of energy or material flows) of producing a given good or service, or of maintaining a given ecological state. Approaches based on physical costs often rely on the first and second principles of thermodynamics and systems ecology approaches. In this case, quantifying turns from monetary valuation to physical accounting. Some pioneer examples within this approach can be traced back to the work of authors such as Podolinsky in the 19th century and Frederick Soddy at the beginning of the 20th century,[62] and later in some ecological economic precursors such as Odum[63] or Georgescu-Roegen.[2] Herewith we will consider two sets of methods: 'surface and material-based methods' and 'energy-based methods'.

4.2.1 Surface or Material Accounting Methods

These methods, which often rely on industrial ecology approaches, quantify flows of materials or surface requirements resulting from socio-economic metabolism.[64,65] The most widely used surface accounting method is the Ecological Footprint,[66,67] which measures the biologically productive land an individual population or activity uses to produce all the resources it consumes and to absorb the wastes it generates (water footprint and carbon footprint methods use similar approaches). Another method that may be classified within this group is Land Cover Flow analysis, which can be used to monitor changes in natural capital quality and land multi-functionality.[68]

Examples of the main materials-based methods are: Material Flow Analysis,[69] which accounts for environmental inputs and outputs in the metabolism of social–ecological systems, and Life Cycle Analysis.[17]

4.2.2 Energy Based Methods

These methods aim to quantify the amount of energy or exergy that needs to be invested to perform any (*e.g.* economic) process. In the first case (energy costs), the main reference is the Embodied Energy Analysis method.[70,71] In the latter, the Exegetic Replacement Cost aims to quantify the costs of non dissipated (that is, useful) energy that natural capital consumption entails.[72,73] The so-called Emergy Synthesis[74,75] is a systems ecology-based method that has capacity to distinguish between different qualities of energy throughout the productive processes being analysed. Finally, Human Appropriation of Net Primary Production (HANPP)[76] can be defined as the difference between the Net Primary Production of the potential natural vegetation and the amount of Net Primary Production remaining in ecosystems.

5   Discussion and Conclusions

5.1   The Controversy of Value Commensurability

The search of a common measuring rod for 'value' has often been the aim pursued in order to clarify value theory. Classical economists such as Ricardo and Marx tried to identify the common substance of value in *labour* (*abstract labour force* in the case of the latter). Some authors from the natural sciences sought to find it in *energy* or any of its derivates, such as exergy. Finally, neoclassical economists tried to find it in the *utility* concept, assuming the measurability of this abstract category in money. All of them aimed to find a mono-value theory.

However, mono-value theories have often been branded to represent different forms of reductionism, being considered to capture only one of value's many dimensions.[10,77] In this context, some authors have suggested the existence of a plurality of values or multiple-value dimensions. In fact within inter- or trans-disciplinary sciences such as ecological economics, there is an increasing recognition about the existence of different forms of value (*e.g.* monetary, ecological and cultural) that cannot necessarily be reduced to a single measuring rod, that is, the existence of *incommensurable* values.[78] However, value incommensurability does not hinder the fact that different alternatives can be compared upon a rational basis. The aggregation of different incommensurable value forms with the aim of their incorporation into a decision-making processes can be operationalised through, for example, multi-criteria evaluation methods.[79]

5.2   Why Use the Notion of Natural Capital?

There is an increasing consensus between environmental and ecological economists that ecological problems are caused, to a large extent, by the persistence of an economic system that is blind to the ecological (and social) deterioration and injustice involved in economic activity. Environmental economists usually refer to these hidden costs as 'externalities',[80] whereas ecological economists often prefer the concept of 'cost-shifting gains'.[81]

In this context, natural capital and ecosystem services are essential conceptual tools that reflect the role ecosystems play in human well-being, not only when they are subject to extraction or active use (and thus marketable), but also when it means to 'just' preserve ecosystems to provide 'free' benefits from, for example, regulating services.

Although the potential that ecosystem valuation and accounting methods may have for re-orienting decision making into a more sustainable pathway has been widely acknowledged in the ecological economic literature, several conceptual and methodological problems still need to be resolved. At the conceptual level, it is worth mentioning the controversies around value commensurability referred above, the commodification of nature,[82,83] and the way valuation processes can act as 'value articulating institutions' shaping the way humans perceive nature according to particular world views and value systems.[55]

If used, valuation of ecosystems and their services should not be understood as an *end* in itself, but rather as a pragmatic tool that aims to incorporate the true contribution of nature to human well-being into economic theory and practical decision making. The conceptualisation of nature in terms of natural capital and ecosystem services does not aim to substitute nature's intrinsic value as the main basis for its conservation or wise use. Instead, it complements ethic-based arguments so that both forms of value can be used synergistically in the search for a sustainable human-nature relationship. In other words, economic valuation of natural capital searches to 'push' conservation arguments beyond conservationist circles and structurally integrate them into every-day decision making (individual, corporate and political) at all scales. A key challenge in this respect is to identify in which situations intrinsic and instrumental values can be used complementarily, and in which ones the use of utility-based rationalities (*e.g.* financial incentives) may undermine moral sentiments for conservation.[84] For example, articulating the wise use of nature around the concepts of natural capital and ecosystem services may be appropriate in a market economic setting, but can be misleading in peasant, indigenous or other community based societies where concepts such as nature and well-being rely on completely different ontologies, and where environmental values are deeply interwoven with community and spiritual values. In such contexts, utilitarian framing of conservation concerns can be misleading and culturally offensive.[85] Thus, the use of economic valuation should take due account of the ecological and socio-economic context, and of the interests of all stakeholders involved in, and affected by, the decision.

References

1. F. Soddy, *Wealth, Virtual Wealth and Debt,* Allen and Unwin, London, 1926.
2. N. Georgescu-Roegen, *The Entropy Law and the Economic Process,* Harward University Press, London, 1971.
3. H. E. Daly, *Beyond Growth,* Beacon Press, Boston, 1996.
4. J. Martínez-Alier, *Ecol. Polít., 2009,* **36**, 23.
5. R. Costanza and H. Daly, *Conservat. Biol.,* 1992, **6**, 37.
6. P. Ekins, S. Simon, L. Dutsch, C. Folke and R. de Groot, *Ecol. Econ.,* 2003, **44**, 165.

7. G. Daily, *Nature's Services: Societal Dependence on Natural Ecosystems,* Island Press, Washington DC, 1997.
8. R. S. de Groot, M. Wilson and R. Boumans, *Ecol. Econ.,* 2002, **41**, 393.
9. Millennium Ecosystem Assessment, *Ecosystems and Human Well-being. A Framework for Assessment,* Island Press, Washington DC, 2003.
10. J. Martínez Alier, *Ecological Economics,* Blackwell, Oxford, 1987.
11. J. M. Naredo, *La Economía en Evolución: Historia y Perspectivas de las Características Básicas del Pensamiento Económico,* Siglo XXI de España Editores, Madrid, 2003.
12. J. L. Weber, *Ecol. Econ.,* 2007, **61**, 695.
13. P. Bartelmus, *Ecol. Econ.,* 2009, **68**, 1850.
14. B. Fisher, R. K. Turner and P. Morling, *Ecol. Econ.,* 2009, **68**, 643.
15. P. R. Armsworth, K. Chan, M. A. Chan, G. C. Daily, C. Kremen, T. H. Ricketts and M. A. Sanjayan, *Conservat. Biol.,* 2007, **21**, 1383.
16. G. C. Daily and P. A. Matson, *Proc. Natl. Acad. Sci. U. S. A.,* 2008, **105**, 9455.
17. O. Carpintero, *El Metabolismo de la Economía Española. Recursos Naturales y Huella Ecológica (1955–2000),* Fundación César Manrique, Islas Canarias, 2005.
18. R. U. Ayres and J. van den Bergh, *Ecol. Econ.,* 2005, **55**, 96.
19. S. Dinda, *Ecol. Econ.,* 2004, **49**, 431.
20. G. M. Grossman and A. B. Kruger, *Q. J. Econ.,* 1995, **110**, 353.
21. D. I. Stern, *World Development,* 2004, **32**, 1419.
22. R. Cleveland, *J. Ind. Ecol.,* 1999, **2**, 15.
23. N. Georgescu-Roegen, in *Scarcity and Growth Reconsidered,* ed. V. K. Smith, John Hopkins University Press, Baltimore, 1979, p. 95.
24. H. E. Daly, *Ecol. Econ.,* 1997, **22**, 261.
25. H. Barnet and C. Morse, *Scarcity and Growth,* Johns Hopkins-Press. Baltimore, 1963.
26. R. M. Solow, *Am. Econ. Rev.,* 1974, **64**, 1.
27. J. M. Naredo, *Raíces Económicas del Deterioro Ecológico y Social,* Siglo XXI de España Editores, Madrid, Spain, 2006.
28. A. Balmford, A. Bruner, P. Cooper, R. Costanza, S. Farber, R. E. Green, M. Jenkins, P. Jefferiss, V. Jessamy, J. Madden, K. Munro, N. Myers, S. Naeem, J. Paavola, M. Rayment, S. Rosendo, J. Roughgarden, K. Trumper and R. K. Turner, *Science,* 2002, **297**, 950.
29. R. Inglehart, *Culture Shift in Advanced Industrial Societies,* Princeton University Press, Princeton, USA, 1990.
30. K. Hubacek and J. van der Bergh, *Ecol. Econ.,* 2006, **56**, 5.
31. E. F. Schumacher, *Small is Beautiful: Economics as if People Mattered,* Harper & Row, New York, USA, 1975.
32. D. Pearce and R. Turner, *Economics of Natural Resources and the Environment,* John Hopkins University Press, Baltimore, USA, 1990.
33. A. M. Jansson, M. Hammer, C. Folke and R. Costanza, *Investing in Natural Capital: The Ecological Economics Approach to Sustainability,* Island Press, Washington, D.C., USA, 1994.
34. J. Boyd and S. Banzhaf, *Ecol. Econ.,* 2007, **63**, 616.

35. K. J. Wallace, *Biol. Conservat.*, 2007, **139**, 235.
36. B. Fisher, R. K. Turner and P. Morling, *Ecol. Econ.*, 2009, **68**, 643.
37. R. Costanza, *Biol. Conservat.*, 2008, **141**, 350.
38. EC (European Commission), *The Economics of Ecosystems and Biodiversity*, European Commission, Brussels, Belgium, 2008.
39. R. Haines-Young and M. Potschin, in *Ecosystem Ecology: a New Synthesis*, ed. D. Raffaelli and C. Frid, BES Ecological Reviews Series, Cambridge University Press, UK, 2010, p.110.
40. E. Gómez-Baggethun and R. de Groot, *Ecosistemas*, 2007, **16**, 4.
41. B. Martín-López, E. Gómez-Baggethun, J. González, P. Lomas and C. Montes, in: *Handbook of Nature Conservation*, ed. J. B. Aronoff, Nova Science Publishers, New York, USA, 2009, p. 261.
42. A. Balmford, A. Rodrigues, M. Walpole, P. ten Brink, M. Kettunen, L. Braat and R. de Groot, *Review on the Economics of Biodiversity Loss: Scoping the Science,* European Commission, 2008.
43. C. S. Holling, *Annu. Rev. Ecol. Syst.*, 1973, **4**, 1.
44. C. S. Holling, *Ecosystems*, 2001, **4**, 390.
45. C. Folke, S. Carpenter, T. Elmqvist, L. Gunderson, C. S. Holling, B. Walker, J. Bengtsson, F. Berkes, J. Colding, K. Dannell, M. Falkenmark, L. Gordon, R. Kasperson, N. Kautsky, A. Kinzig, S. Levin, K. G. Mäler, F. Moberg, L. Ohlsson, E. Ostrom, W. Reid, J. Rockström, H. Savenije and U. Svedin, *Resilience and Sustainable Development: Building Adaptive Capacity in a World of Transformations*, Scientific Background Paper on Resilience for the process of The World Summit on Sustainable Development on Behalf of The Environmental Advisory Council to the Swedish Government, 2002.
46. S. R. Carpenter, B. Walker, J. M. Anderies and N. Abel, *Ecosystems*, 2001, **4**, 765.
47. C. Folke, S. Carpenter, B. Walker, M. Scheffer, T. Elmqvist, L. Gunderson and C. S. Holling, *Annu. Rev. Ecol. Syst.*, 2004, **35**, 557.
48. R. de Groot, *Landscape Urban Plan.*, 2006, **75**, 175.
49. R. S de Groot, *Functions of Nature: Evaluation of Nature in Environmental Planning, Management and Decision Making,* Wolters-Noordhoff BV, Groningen, The Netherlands, 1992.
50. R. Costanza, R. d'Arge, R. de Groot, S. Farber, M. Grasso, B. Hannon, K. Limburg, S. Naeem, R.V. O'Neill, J. Paruelo, G. R. Raskin, P. Sutton and M. van der Belt, *Nature*, 1997, **387**, 253.
51. G. C. Daily, T. Soderquist, S. Aniyar, K. Arrow, P. Dasgupta, P. R. Ehrlich, C. Folke, A. Jannson, B. O. Jansson, N. Kautsky, S. Levin, J. Lubchenco, K.-G. Mäler, S. David, D. Starrett, D. Tilman and B. Walker, *Science*, 2000, **289**, 395.
52. J. Martínez Alier, *The Environmentalism of the Poor: a Study of Ecological Conflicts and Valuation,* Edward Elgar, Cheltenham, UK, 2002.
53. L. Pritchard, C. Folke and L. Gunderson, *Ecosystems*, 2000, **3**, 36.
54. B. Martín-López, C. Montes and J. Benayas, *Environ. Conservat.*, 2007, **34**, 215.

55. A. Vatn, *Institutions and the Environment*, Edward Elgar, Cheltenham, 2005.
56. A. Straton, *Ecol. Econ.*, 2006, **56**, 402.
57. D. W. Pearce and R. Turner, *Economics of Natural Resources and the Environment*, John Hopkins University Press, Baltimore, 1990.
58. Y. E. Chee, *Biol. Conservat.*, 2004, **120**, 459.
59. C. Spash, *Environ. Values*, 2000, **9**, 453.
60. A. Chiesura and R. de Groot, *Ecol. Econ.*, 2003, **44**, 219.
61. M. Kumar and P. Kumar, *Ecol. Econ.*, 2008, **64**, 808.
62. J. Martínez Alier, *Los Principios de la Economía Ecológica. Textos de P. Geddes, S. Podolinsky y F. Soddy*, Fundación Argentaria, Madrid, Spain, 1995.
63. H. T. Odum, *Environment, Power and Society*, John Wiley and Sons, New York, USA, 1971.
64. M. Patterson, *Ecol. Econ.*, 1998, **25**, 105.
65. H. Schandl, C. M. Grünbühel, H. Haberl and H. Weisz, *Handbook of Physical Accounting. Measuring Bio-Physical Dimensions of Socio-Economic Activities: MFA–EFA–H'ANPP*, Social Ecology Working Paper, Vienna, July 2002, no 73.
66. M. Wackernagel and W. E. Rees, *Ecol. Econ.*, 1997, **20**, 3.
67. M. Wackernagel, L. Onisto, P. Bello, A. Callejas Linares, I. S. López Falfán, J. Méndez García, A. I. Suárez Guerrero and M. G. Suárez Guerrero, *Ecol. Econ.*, 1999, **29**, 375.
68. R. Haines-Young and J. L. Weber, Land Accounts for Europe 1990–2000, Towards Integrated Land and Ecosystem Accounting, EEA Report, Copenhagen, 2006, no 11.
69. P. L. Daniels and S. Moore, *Ind. Ecol.*, 2002, **5**, 69.
70. P. Chapman, *Energy Policy*, 1974, **2**, 91.
71. R. Costanza, *Science*, 1980, **210**, 1219.
72. J. M. Naredo and A. Valero, *Desarrollo Económico y Deterioro Ecológico*, Fundación Argentaria-Visor, Madrid, Spain, 1999.
73. J. M. Naredo, in: *The Sustainability of Long Term Growth: Socioeconomic and Ecological Perspectives*, ed. M. Munasinghe and O. Sunkel, Edward Elgar Publishing, Cheltenham, Northampton, UK, 2001.
74. H. T. Odum, *Environmental Accounting: Emergy and Decision Making*, John Wiley, New York, USA, 1996.
75. H. T. Odum, *Energy Accounting*, John Wiley, New York, USA, 2000.
76. P. M. Vitousek, P. Ehrlich, A. Ehrlich and P. Matson, *BioScience*, 1986, **34**, 368.
77. N. Georgescu-Roegen, *El Trimestre Económico*, 1983, **198**, 829.
78. J. Martínez Alier, G. Munda and J. O'Neill, *Ecol. Econ.*, 1998, **26**, 277.
79. G. Munda, *Eur. J. Operat. Res.*, 2004, **158**, 662.
80. A. C. Pigou, *The Economics of Welfare*, Cossimo Classics, New York, USA, 2006.
81. W. Kapp, in: *Social Costs, Economic Development, and Environmental Disruption*, ed. J.E. Ullmann, University Press of America, Lanham, USA, 1983.

82. D. J. McCauley, *Nature*, 2006, **443**, 27.
83. C. Spash, *Environ. Values*, 2008, **17**, 259.
84. S. Bowles, *Science*, 2008, **320**, 1605.
85. K. T. Turner, J. Paavola, P. Cooper, S. Farber, V. Jessamy and S. Georgiu, *Ecol. Econ.*, 2003, **46**, 493.

Protecting Water Resources and Health by Protecting the Environment: A Case Study

LUKE DE VIAL, FIONA BOWLES AND P. JULIAN DENNIS

ABSTRACT

The effluent from growing populations and the increasing use and disposal of chemicals into the environment can overwhelm ecosystems' natural ability to absorb and treat waste. Historically, the approach to protecting the environment from sewage discharged into it is to collect and treat it. Similarly water taken from sources used for public water supply is treated to protect populations from the effects of untreated or diffuse pollution. Both approaches have proved successful and will continue to play an essential role in protecting ecosystems and public health. However, if the cost of treatment and the associated carbon footprint of these treatment processes are to be controlled, a complementary approach is needed to protecting ecosystems and the water extracted from them for public supply. Catchment management, a means of minimising diffuse pollution, and therefore the need for more complex treatment, is being used by Wessex Water and other water companies to drive down the cost and carbon footprint of more expensive and energy-intensive treatment processes needed to meet the demands of tightening health and environmental standards. The impact of this approach in a number of defined catchments in the Wessex Water region has been studied over the last three years. Whilst the catchment management approach has demonstrated that a reduction in the residual levels of nitrogen in the soil can be achieved without affecting a farm's output, it is not yet clear whether the impact of these changes has reached the water supply aquifer. However, in surface waters where travel times are much faster, immediate impacts have been achieved, particularly a reduction in the detection of pesticides.

Issues in Environmental Science and Technology, 30
Ecosystem Services
Edited by R.E. Hester and R.M. Harrison
© Royal Society of Chemistry 2010
Published by the Royal Society of Chemistry, www.rsc.org

1 Introduction

Ensuring populations have clean fresh water to drink is an essential function of water utilities today. This need is nothing new. Two thousand years ago, Roman engineers built aqueducts to bring water into cities where population growth and density had exhausted or polluted local water sources beyond use. It was recognised even then that clean water was essential in maintaining healthy vibrant populations.

Even so, little thought was given to the disposal of waste products that might affect health or the environment.[1] With a sufficiently low population and relatively little industry, the aquatic environment has an ability to dilute, absorb or treat organic waste. Thus, the effects of effluent disposal for small populations went unnoticed, whilst those for the large growing cities, such as London, Manchester and Liverpool, were simply ignored. It therefore became practice to simply dilute and disperse the effluent into the environment and allow it to be treated naturally. In many respects this worked well, particularly for marine outfalls. Eventually, however, with growing populations and additional diverse sources of pollution, natural ecosystems with their capacity to deal with pollutants could no longer cope with the growing quantity and type of inputs, and were simply overwhelmed.

Dr John Snow confirmed the link between water contaminated with polluted river water and water-borne disease.[1] Thus water-borne pathogens and the need to control water-borne disease epidemics were the key drivers for the development of water supply and treatment practices of the 19th and 20th centuries. It was understood that taking effluent and disposing of it in rivers down stream of water intakes helped prevent disease. Furthermore, simple treatments, such as filtration through sand and the addition of chlorine (or both), helped to make water safe to drink.[1]

The water industry is unusual, however, with respect to the fact that it discharges its waste to the water environment from which it extracts its raw material for drinking water supply. Thus locating the abstraction points for public water supply upstream of sewage discharges became standard practice. Population growth and the increasing variety and use of chemicals that find their way into the waste streams of domestic and commercial practice has led to the need for ever more sophisticated and energy-intense treatment to make effluent less polluting before disposal.

In addition, chemicals that are lost to or used in the environment, particularly chemical fertilisers, herbicides and pesticides required to provide the quantity of food needed by growing populations, and other pollutants from roads and pavements *etc.*, also find their way into water sources. To meet the drinking water quality standards set for water supply[2] these pollutants have to be removed.

Therefore to protect the environment and water resources, current practice relies on treating sewage effluent at the end of the sewerage system, in the sewage treatment works (end of pipe treatment). Then further treatment of the water at the water treatment works, abstracted for public supply, to remove chemicals present as a result of pollution. The cost and consequence of dealing with these wastes is not paid for by the polluter or many of the retailers that benefit from current industrial and agricultural practice, but by water consumers through their utility bills and ultimately by the environment.

The recognition of the value and therefore growing concern for the environment, rather than just for the human population, has been enshrined in national and European legislation.[3–5] The water industry has had to develop better treatment techniques for both drinking water[3] and sewage[5] and address the environmental impacts on water resources. Since privatisation in 1990, which was designed to get more investment into the water sector to meet European standards for drinking water and effluent discharged to the environment, customers bills have risen by approximately 50% in real terms.

At the Second World Water Forum in The Hague in 2000,[6] the declaration highlighted the role of sustainable water resource management in ensuring the integrity of ecosystems. The protection of the water ecosystems themselves was enshrined in Europe in the 2000 Water Framework Directive (WFD).[7]

As a result of these new regulations, standards have risen over the past 30 years for both drinking water quality and effluent discharged, whilst protection of the environment, which is inevitably affected by these changes, has also increased. Specific duties to the water industry, as a statutory undertaker or competent authority, are included within the Habitats Regulations 1994,[8] Countryside and Rights of Way Act 2000,[9] The Environmental Assessment of Plans and Programmes Regulations 2004[10] and Natural Environment and Rural Communities Act 2006.[11] These legal duties are included in the prescriptive methodologies for developing water resource management plans and water company-regulated business plans. Thus assessment of environmental impact and monetisation of the effect on ecosystems is already part of strategic planning activities within the water industry and has been a major driver in the increase in water and sewerage bills since privatisation in 1990.

2   The Environmental Obligations on Water Utilities

Water companies work with the water environment. They abstract water from the environment and therefore are influenced greatly by the quality and quantity of the available water, and they discharge treated sewage effluent back into the environment and must do so in a manner that does not cause harm.

These processes are heavily regulated. Abstraction of more than $20\,\text{m}^3\,\text{day}^{-1}$ requires a licence from the Environment Agency (EA).[12] Abstraction licences, particularly those granted in the last 20 years, typically have many environmental flow conditions attached. Some licences restrict or prohibit abstraction when river flows are low; others have considerable environmental monitoring conditions attached.

To protect public health water companies must then treat the water. The treatment standards are laid out in the drinking water standards.[2] The raw

Table 1 Drinking water standards commonly exceeded in raw water.

Determinand	Drinking water standard
Nitrate	$50\,mg\,l^{-1}$
Individual pesticide	$100\,ng\,l^{-1}$
Total pesticides	$500\,ng\,l^{-1}$
Cryptosporidium	Risk assessed – no significant presence
Coliforms	Absence

water from the environment commonly exceeds the standards set out in Table 1.

Compliance with the drinking water standards is regulated by the Drinking Water Inspectorate (DWI).

A discharge of treated sewage effluent requires a discharge consent.[5,12] The purpose of the consent is to protect the environment, particularly the water into which the effluent is discharged. The consent will specify a flow rate and water quality conditions, setting permissible levels of suspended solids, biochemical oxygen demand (BOD) and ammoniacal nitrogen.

The regulatory standards that ensure environmental protection are not static. Over the last twenty years there has been a considerable tightening of the standards. For example in the area of abstraction, Wessex Water has made reductions of approximately $30\,Ml\,d^{-1}$ in licensed abstractions since privatisation to meet environmental concerns and a further $23.5\,Ml\,d^{-1}$ is planned over the next 10 years.

Changes to the drinking water standards have resulted in significant investments. The lead standard has been reduced from 50 to $25\,\mu g\,l^{-1}$.[2] In 2013, this standard will be further reduced to $10\,\mu g\,l^{-1}$. In addition, new standards have been introduced for turbidity and cryptosporidium which have triggered major investments.

2.1 Discharge Consents have Tightened

The move towards improving water quality, for compliance with the EU Freshwater Fish Directive[4] and tighter river water quality objectives in preparation for the WFD, has resulted in reviews of discharge consents over the last ten years. This has been particularly noticeable at small rural sewage treatment works that discharge into the headwaters of watercourses, where little dilution is available. The resultant consents have required high levels of treatment which often require energy-intensive tertiary treatment to meet these new standards.

2.2 Review of Discharge Consents due to EU Directives

The concern over coastal water quality and in particular the influence of treated sewage discharges on bathing and shellfish waters[13] has driven a major expansion of treatment at coastal sites. Improvements to sewerage systems have been necessary to reduce the impact of intermittent storm water discharges. Continuous treated waste water discharge now receives disinfection either by ultraviolet disinfection or by use of membrane-treatment technology.

Many inland rivers have now been designated as 'Sensitive Waters'[5,12] and phosphate-reduction processes have been installed, usually using iron salts to precipitate the phosphate and reduce nutrient load to receiving water courses.

2.3 Prohibition of 'Dumping' Sewage at Sea

The practice of marine disposal ceased in 1998. Sludge had been tankered out to marine areas of high dilution and dispersion and discharged to sea. Since 1998, this sludge has been treated with lime or anaerobic digestion to meet the appropriate bacteriological standards for recycling and is taken to farmland and used as a soil conditioner.

Even tighter regulation is proposed, in particular the WFD[7] aims to protect all elements of the water cycle and enhance the quality of aquifers, rivers, lakes estuaries and the sea. Its emphasis is on ecological status, with reference to pristine exemplars although it recognises that some watercourses have been too heavily modified for use to achieve the same high ecological status, and a high ecological potential instead will be targeted. The WFD requires good ecological status to be achieved by 2015, but allows derogations, in particular for measures required which are 'disproportionately expensive'. In the UK in 2009, a series of River Basin Management Plans were published[14] which identify the current status of each river basin, or group of catchments, with a programme of measures by which good ecological status (or potential) will be achieved. In these first River Basin Management Plans, the greater proportion of measures (and costs) falls upon the water industry. The total cost of the WFD to the South West from 2009 to 2015 will be £2700 million, with approximately 96% of this cost falling to the Water Industry and our customers.

Eighty per cent of the Water Industry National Environmental Programme (NEP)[15] for AMP5 (the Asset Management Plan period that runs between 2010–2015) is required to meet the WFD standards (EA personal communication). These schemes include environmental investigations, which may lead to further investment post-2015, and physical changes that are required to our abstractions or discharges in the next five years.

Drinking water quality standards are not expected to change significantly in the future. However, there are a number of emerging threats such as metal-dehyde, clopyralid and perfluorooctanate sulfonate (P.F.O.S) problems. Commonly the water industry only becomes aware of these as potential problems in the raw water supplies as analytical methods are developed to detect them in very low concentrations. However, once we are aware that such pollutants are present they must be removed from drinking water to remain compliant with the standards.

3 How Water Utilities meet their Environmental Obligations

Water companies are heavily regulated private sector utilities and are very concerned about meeting their environmental obligations, in particular the conditions specified in abstraction licences, discharge consents and drinking water standards.

The UK water industry was privatised in 1990 to provide the management and capital to ensure that environmental standards were met. The improvement in the last twenty years has been impressive. In Wessex Water alone:

- The disposal of sewage sludge at sea has stopped.
- Compliance with abstraction licences has improved markedly.
- 'Unacceptable' abstraction licences of 30 Ml d^{-1} have been reduced.
- Compliance with the drinking water standards has gone up from 98% to 99.98%.
- Compliance with discharge consents has gone up from 96.7% to 99.8%.
- Secondary treatment has been provided at sites with a population equivalent of over a million. Tertiary treatment has been installed to some 30 sites.
- Over 70 first-time rural sewerage systems have been provided to overcome environmental impacts.

The improvements above relate to achieving environmental standards and improving coastal water quality. In addition visible environmental improvements have been achieved, such as rivers that no longer dry out during dry years (Malmesbury Avon) and salmon returning to previously polluted rivers.

These improvements have been achieved largely by engineering works, primarily the installation of additional treatment and sewage works. In many cases, the reason for this was fairly straightforward, *i.e.* a 'treatment-type' solution was the only credible solution. This would have been the case for most secondary treatment at sewage works.

However, other more complex reasons for this engineering-based approach include a perceived 'engineering bias' in water companies; the lack of control of pollution at source or an 'integrated' catchment approach; the lack of funding for alternative approaches; and the different treatment of capital and operating expenditure by the water industry's economic regulator.

4 A More Sustainable Ecosystem-Based Approach for the Future

This engineering approach is, however, reaching its viability limits in many areas.

Gross problems in terms of water supply treatment and sewage work discharges have in most cases been overcome. The problems we are now trying to solve have much smaller marginal benefits. For instance, on some rivers we are now trying to reduce the impact of abstraction on river flows from a current level of 20% of natural flow down to an impact of no greater than 10%. This contrasts with previous improvements when abstraction was taking 100% of flow under certain conditions. Similarly on the waste water side all sewage works now have at least secondary treatment, the debate now is whether tertiary treatment (sand filtration of treated effluent or disinfection) should be provided. The cost of the schemes and the reduced benefits relative to previous improvements is making the cost/benefit of such schemes difficult to justify.

High cost also goes alongside a high carbon foot-print both to build and operate. Like all sectors, water companies are trying to reduce their carbon impact through such measures as electricity generation using methane from sludge. However, building and operating new energy-intensive processes limits the overall progress that can be made in carbon reduction.

In some cases attaining the standards by an engineering approach is becoming impractical. A recent example of this is the detection of metaldehyde in raw water supplies. Metaldehyde is the active ingredient in the most commonly used slug pellets. In the last two years an analytical method for its detection at levels close to the drinking water standard $(100\,\mathrm{ng}\,\mathrm{l}^{-1})$ has been developed. Testing has shown it to be present in many raw surface water supplies across the country. It is, however, very difficult to remove metaldehyde from water. The treatment processes which have been suggested as being effective are very capital and energy intensive.

A further limit on continuing with this expensive engineering approach is the level of water company bills. To deal with the environmental improvements required, water prices have increased by approximately 50% in real terms since privatisation. Whilst water bills generally remain below the level of gas and electricity bills, the number of people in water poverty (defined as those spending more than 3% of their disposable income on water) has increased since privatisation.

As a consequence, water companies have started to try to take an alternative integrated catchment ecosystem based approach, sometimes counter to the regulatory and financial orthodoxy.

The best example of this has been the approach taken by a number of companies in trying to prevent the pollution of water supplies at source, rather than by providing additional treatment. Strictly speaking this is the domain of landowners and the Environment Agency. Nevertheless some water companies have felt it necessary to become involved.

This 'catchment management' approach has involved water companies working with farmers to change practices to reduce farm pollution levels. The methods used include data, advice, practical help and, in some circumstances, financial incentives.

The arguments for this approach are very simple: lower cost, lower carbon, wider ecosystem benefits and social engagement.

This approach has been supported by regulators particularly the DWI and, more recently, by Ofwat. The European Commission and Defra have also supported the approach, not least in their funding of the WAgriCo project.[16]

5   The Wessex Water Experience with Catchment Management

Wessex Water has been trying to address a nitrate problem, *via* catchment management, in the raw water of groundwater sources and pesticides in the raw water of both groundwater and surface water sources (reservoirs).

As regulated businesses, water companies produce business plans in a five year cycle. At the beginning of the last cycle in 2005, Wessex Water took the view that the nitrate problem was so bad at two groundwater sources that nitrate treatment was required as soon as possible. At four other sites affected by nitrate and three groundwater sources at risk from pesticides we have taken a catchment-based approach. Figure 1 illustrates the problem.

Catchment management has involved a range of activities with farmers. These activities have been focussed in the catchment of the water supply sites. The catchment approach has involved the following steps:

- Identification of the catchment of the water supply source and farming activities within this catchment.
- Actions taken with the farmer to reduce pollution to the source (and therefore to the wider environment).
- Monitoring of improvements.

5.1   Identification of Catchment and Farms

The first phase of the work is to identify the area over which improvements are sought. For surface water sources, this is the surface water catchment upstream of the abstraction point. For groundwater sources, identification is more difficult. Initially, Wessex Water used the published EA groundwater protection zones. Part of the Wessex Water work involved the construction of additional boreholes in the catchment to monitor pollution levels, but water level data from these boreholes could also be used to refine the definition of the groundwater catchment. In the case of the Area B source near Dorchester this led to a radical redefinition of the catchment area.

The second phase of the work was to identify the farmers, who they are and what crops they grow and whether their activities are likely to be contributing to the pollution of the catchment.

5.2   Actions Taken with the Farmer to Reduce Pollution

Once the catchment management boundaries were defined, the next task of the catchment management project was to identify the main issues affecting, or potentially affecting, nitrate levels in the catchment. Initially, a catchment walk-over survey using footpaths and byways was conducted to identify any obvious issues before contacting the landowners. The farmers were then contacted individually and visited to explain Wessex Water's nitrate issues and to raise their general awareness of practices that might affect nitrates in

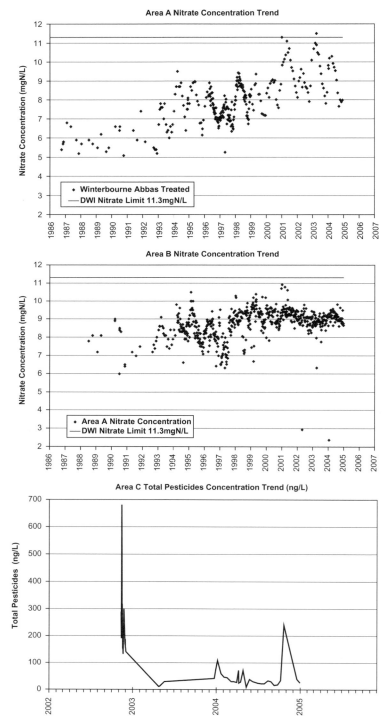

Figure 1   Water quality problems in Area A, Area B and Area C, up to 2005.

groundwater. The initial phase enabled the identification of the areas of concern.

These included:

- Variable standards of farming practices.
- Presence of intensive dairy units.
- Farming of outdoor pigs in close proximity to Area B source.
- High levels of nitrates in boreholes in the dairy yard.
- Thinness of soils.
- Lack of slurry or dirty water storage.
- High rates of fertiliser use.
- Poor maintenance and calibration of fertiliser and manure application equipment.
- Long-term storage of manures in the same place on chalk aquifers close to water sources.
- Private boreholes situated in dairy yards.
- Age of equipment.
- Areas of continuous maize growing.
- Unlined slurry and silage pits.

Some of these practices are illustrated in Figure 2.

More detailed analysis of farming practices and techniques allowed the categorisation of farms into low, medium or high risk.

Many of the on farm issues have been addressed by close and regular contact with the farmers. This has allowed Wessex Water's catchment advisers to suggest alterations and modifications to farm practices. These include measures such as:

- Reducing inorganic fertiliser use by taking into consideration manure and soil mineral nitrogen supply.
- Calibration of fertiliser and manure spreaders.
- Soil mineral nitrate sampling, potash and phosphate soil sampling so as to tailor fertiliser rates to crop requirements.
- Improvement in soil management *via* soil management plans to maximise crop uptake of soil nutrients.
- Adoption of resource protection measures under Environmental Stewardship.
- Change in cultivations.
- Altering drilling dates of autumn sown crops.
- The use of catch crops (a winter crop that utilises remaining surplus nutrients in the soil, which is then ploughed under in spring prior to sowing a spring crop).
- Analysis of manures.
- Nutrient and manure management plans.
- Providing data on water quality through monitoring streams, private boreholes and public sources.

Figure 2 Area B catchment: examples of farming practices that can influence nitrate levels in groundwater.

5.3 Monitoring of Improvements

The effect of the catchment management work can be traced through the farming and hydrological cycle. This is illustrated in Figure 3 for nitrates, although the same principles would apply to pesticides. This assessment starts at the farm in terms of nitrate use and then sequentially through the nitrate levels seen in the soil, in the water leaching out of the soil, in the rivers and ground waters of the catchment and finally at the public water supply source.

Some of the results from this monitoring are shown in Figure 4. They show nitrate application rates, the amount of nitrate residue in the soil after harvest and a calculation of how much nitrate will have leached from the soil. These graphs indicate an improvement over time, since the catchment management work began in 2006.

The nitrate loss from the soil to groundwater can also be measured directly using porous pots. These pots are installed at the base of the root zone and capture water moving downwards. Pots were installed in a number of fields covering differing crops and soil types across the catchment. Water samples are taken fortnightly from the pots and analysed for nitrate.

It is not the slope of the line which is of importance when interpreting these graphs, but the nitrate levels at the start of drainage and the total area under the line which is used to determine total loss of nitrates to groundwater. The porous pots are only operational when there is drainage from the soil. The period for which drainage occurs will vary year on year.

Losses under different crops vary significantly and also from year to year, so with porous pots there is a reasonable amount of experience required for interpretation. The porous pot data is useful in measuring the effect of different cropping patterns and varying agricultural practices on nitrate leaching.

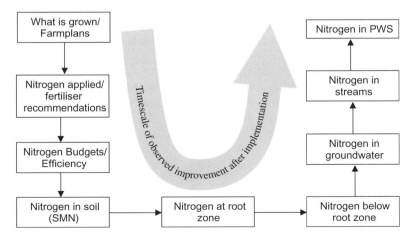

Figure 3 Assessment method for catchment nitrate reduction.

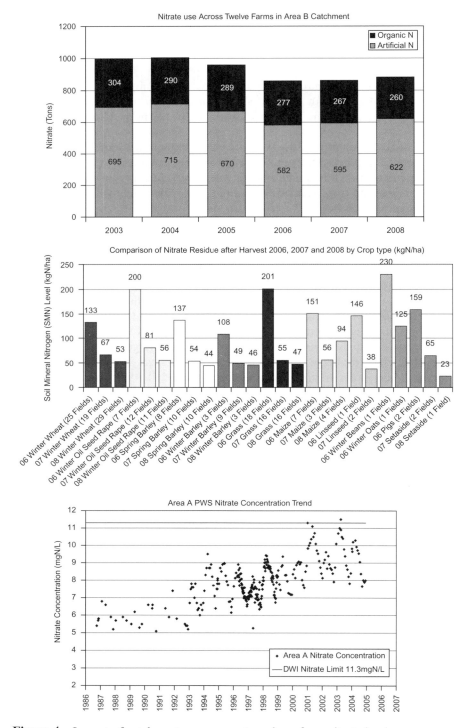

Figure 4 Impact of catchment management work on farm nitrate levels.

For example, stubble turnips or oilseed rape mop up a significant proportion of the available nitrogen, unlike stubbles where there is no crop, or winter wheat where there is little crop over the winter.

The consequences of ploughing out grassland and drilling winter wheat are illustrated in Figure 5, with significant leaching of nitrate from the rooting zone.

It can be seen in Figure 6 that losses, even under the same crop, vary from year to year, but that losses each year are decreasing.

It is interesting to note in Figure 6 that at our Sutton Poyntz site there is no discernable nitrate losses to groundwater. This site is a Site of Special Scientific Interest (SSSI), no fertiliser has ever been applied and it has never been grazed. Under commercial crops it would be impossible to achieve the same results, but by managing the crop rotation we can bring down nitrate losses due to leaching. In this instance we are using cover crops such as stubble turnips to remove excess winter nitrogen.

This data is fed back to the catchment farmers to inform their cropping plans. It must be reiterated that although the porous pot data is very good for showing the variations in nitrate leached from the soil, it is complicated by many variables. It is useful for looking at the nitrate leached from a given field, but year-on-year reductions in leaching are difficult to spot because it is seldom the case (except under permanent grass) that the crop in that field is the same as the previous year.

The concentration of nitrate is also recorded in observation boreholes across the catchment, as well as the public water supply source. Whilst the previous upward trend has been arrested there is as yet no clear downward trend. Also to

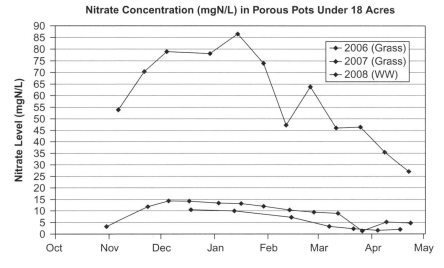

Figure 5 Porous pot data showing the impact of ploughing up grassland on nitrate leaching.

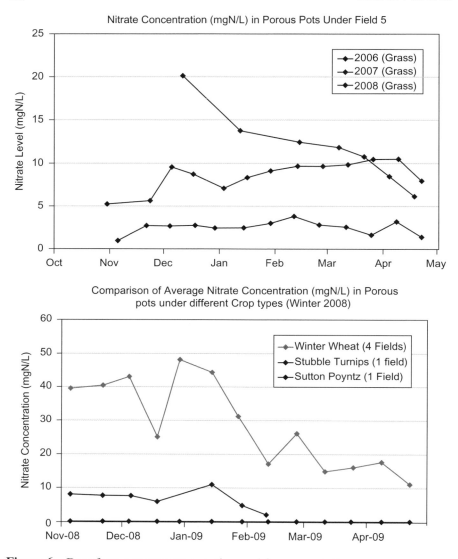

Figure 6 Data from porous pots over time and for crop rotations.

date it has not been possible to distinguish the benefits of the catchment management from other changes and natural variability.

6   Advantages of the Catchment Management Approach

The advantages of the catchment management approach, that it provides relative to the installation of treatment facilities, are lower cost and a lower carbon solution to the problem of poor quality raw water supplies. In addition, this approach offers wider benefits to the ecosystem and involves a widespread community, particularly farmer engagement, in the water pollution issue.

These benefits need to be countered by the increase in risk that the regulatory standards for drinking water may be breached, whilst (and if) the ecosystem approach is proved effective.

The degree to which catchment management is more cost effective than treatment depends on the approach taken to catchment management and the size of the catchment. The more that is spent on catchment management, the less cost effective it will be if, for instance, financial subsidies are involved. However, the more measures deployed, the greater the chance of success. Catchment management is more cost effective in small catchments as the chance of success is greater, focus is easier and the costs of treatment plants are non-linear with size.

Catchment management is a low carbon, almost zero carbon, approach. This compares well with treatment options where there is a considerable carbon impact in building and operating the treatment plant. For instance, the carbon impacts of running a $7\,Ml\,d^{-1}$ treatment plant to remove nitrate are estimated at 1557 tonnes of carbon to build the treatment plant and 27 tonnes of carbon each year to operate it.

Whilst the Wessex Water effort at reducing nitrate and pesticide levels in the catchment is focussed on ensuring the quality of the water abstracted at its sources, there will also be an improvement in the quality of the water in the catchment that is not abstracted.

The social engagement side of catchment management work should not be underestimated, particularly in terms of the longer term protection of water sources. This protection not only applies to nitrates and pesticides, but by raising farmer awareness of the water supply activity in their 'patch', they are more likely to be careful with other potential pollutants, such as diesel stores for example. Many times during the catchment work when individuals were made aware of a problem, the response received was, 'if I had only known, then I would have . . . '.

This is a very real benefit, because whilst ion-exchange plants are good at removing nitrate they provide no protection against pesticides, let alone diesel. Likewise a carbon filter may protect against pesticides and give limited protection against diesel, but it does nothing to reduce nitrate levels. Catchment management should improve both the targeted chemical and reduce the risk of future raw water problems.

These advantages must be balanced with the increased risks. Unlike with treatment, or even blending with other water supplies, there is not a sudden, irrefutable drop in the level of the contaminant in the treated water. As illustrated in the previous figures, whilst after three years' effort the impact of catchment management in groundwater areas may be demonstrable in the quantity of nitrate fertiliser, residual levels of nitrogen in the soil and even the level of nitrate in passing below the root zone of crops, it is far from clear that the impact of these changes has reached the groundwater in the catchment, let alone the abstraction point. By contrast, in surface water systems, where travel times are measured in days, more immediate impacts have been demonstrated.

However, there remains a significant risk to compliance. Wessex Water has tried to mitigate this risk by focussing its activities on effective catchment management initiatives, but also by keeping regulators informed of progress.

7   Other Examples of an Ecosystem Approach

Similar approaches are being taken by other companies to reduce nutrients and colour entering rivers and reservoirs. The work done in recent years has encouraged a large number (39) and a wide range of projects to be proposed for the period 2010 to 2015.

On the waste water side of the business there has been much discussion of trying to control the phosphate in sewage discharges by removing the phosphate at source (*e.g.* washing powder), rather than stripping it out of the water as part of the sewage treatment process.

8   Conclusions

Engineering solutions designed to meet environmental and health requirements are reliable. They are also expensive, however, have a high carbon impact and resolve only particular problems rather than solving and preventing pollution at source. Sometimes engineering solutions are not even available. They also do nothing for the wider environment.

In the last three years Wessex Water and other water utilities have been taking the alternative approach of trying to solve water quality failures due to agricultural pollution at source.

There are signs of progress, particularly in surface water catchments where there are shorter response times to land management changes.

References

1. P. Vinten-Johansen, H. Brody, N. Paneth, S. Rachman and M. Rip, *Cholera, Chloroform and the Science of Medicine, A Life of John Snow,* Oxford University Press, Oxford, UK, 2003.
2. *The Water Supply (Water Quality) Regulation 2000.*
3. *Surface Water for Abstraction Directive* (75/440/EEC).
4. *Fresh Water Fish Directive (78/6/59/EEC* as amended by 2006/44/EC).
5. *Urban Waste Water Directive* (91/271/EEC).
6. *Water for People Campaigns:* The Second World Water Forum in The Hague, Mar 17–33, 2000.
7. *EU Water Framework Directive* (Directive 2000/60/EC).
8. *The Conservation (Natural Habitats, & c.) Regulations 1994*, Statutory Instrument 1994, no. 2716.
9. *Countryside and Rights of Way Act 2000*, ch. 37.

10. *The Environmental Assessment of Plans and Programmes Regulations 2004*, Statutory Instrument 2004, no. 1633.
11. *Natural Environment and Rural Communities Act 2006*, ch. 16.
12. *Water Act 2003, to Amend Water Resources Act 1991*, 2003, ch.37.
13. *European Community Shellfish Directive* (79/923/EEC).
14. *European Water Framework Directive* (2000/60/EC). http://www.environment-agency.gov.uk.
15. *National Environment Programme (NEP) for PR09*. www.environment-agency.gov.uk.
16. *Water Resources Management in Cooperation with Agriculture (WAgriCo)*. http://www.wagrico.org.

Life Cycle Assessment as a Tool for Sustainable Management of Ecosystem Services

ADISA AZAPAGIC

ABSTRACT

Humanity depends on healthy ecosystems: they support or improve our quality of life, and without them, the Earth would be uninhabitable. However, over the past 50 years, fast-growing demands for food, fuel, water and other natural resources have led to an unprecedented degradation of many ecosystem services so that their ability to sustain future generations can no longer be taken for granted. Therefore, reversing ecosystem degradation is one of the great challenges of sustainable development. This is by no means a trivial task as it requires action by all actors in society, including governments, industry and individuals. One of the difficulties is that, even if there was a universal commitment to sustainable development, it is still unclear what goods, services and activities are sustainable and how they could be identified. In an attempt to contribute towards a better understanding of this complex problem, this paper illustrates how environmentally-sustainable products and activities can be identified using Life Cycle Assessment as a tool. Four industrial sectors and supply chains are discussed: energy, transport, industry and food & agriculture. The examples used within these sectors include, respectively, electricity generation; different transportation options and fuels; chemicals and related products; different food products and packaging. Their impacts on the ecosystem services are examined from 'cradle to grave' to help identify more sustainable alternatives. For illustration purposes, the discussion centres on carbon footprint (global

Issues in Environmental Science and Technology, 30
Ecosystem Services
Edited by R.E. Hester and R.M. Harrison
© Royal Society of Chemistry 2010
Published by the Royal Society of Chemistry, www.rsc.org

warming potential), as one of the major global impacts; however, other impacts are also discussed as appropriate.

1 Introduction

Humanity depends on healthy ecosystems: they support or improve our quality of life, and without them, the Earth would be uninhabitable.[1] The Millennium Ecosystem Assessment report (MA)[2] defines ecosystem services as the processes by which the environment produces resources utilised by humans, such as clean air, water, food and materials. Generally, there are four categories of ecosystem services:

- Provisioning services, which include products obtained from ecosystems, including food, fuel, bio-chemicals, medicines and fresh water;
- Supporting services, used for the production of other services including soil formation, photosynthesis, nutrient and water cycling;
- Regulating services, providing benefits from the regulation of ecosystem processes, including air quality, climate, water and disease regulation; and
- Cultural services, providing non-material benefits, such as recreation and aesthetic experiences.

The MA shows that over the past 50 years, humans have changed ecosystems more rapidly and extensively than in any comparable period in human history, largely to meet fast-growing demands for food, fresh water, timber, fibre, and fuel.[2] While these changes have contributed to enormous gains in human well-being and economic development, they have also caused degradation of many ecosystem services. According to the MA, approximately 60% of the ecosystem services are being degraded or used unsustainably, including 70% of provisioning and regulating services.[2] WWF estimate that already in 2005 the world eco-capacity was exceeded by 30% and if the current consumption patterns continue, we will need three planets by 2050 to sustain the growing population numbers.[1]

The bottom line is that human actions are depleting Earth's natural capital, putting such strain on the environment that the ability of the planet's ecosystems to sustain future generations can no longer be taken for granted.[2] At the same time, the MA assessment shows that with appropriate actions it is possible to reverse the degradation of many ecosystem services over the next 50 years, but the changes in policy and practice required are substantial and not currently underway.

It is therefore clear that more sustainable production and consumption patterns must be found if we are to reverse the unsustainable trends. This will require simultaneous consideration of social, environmental and economic dimensions of sustainable development. Currently, we treat these three dimensions as if they are independent of each other, rather than intricately intertwined. This is illustrated in Figure 1: industrial systems draw on the ecosystem services to provide goods and services to society, which are

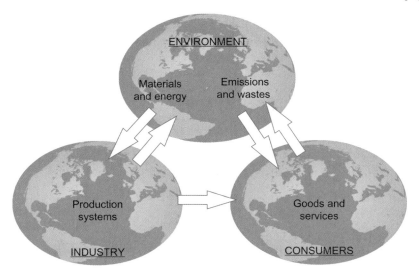

Figure 1 The 'three-planet economy': current management of ecosystems is unsustainable.

eventually discarded as waste back to the environment. All these activities are mainly optimised for economic cost but not necessarily for environmental impacts and better quality of life. This is exactly the challenge of sustainable development: moving from a 'three-planet economy' to a 'one-planet economy' will require that goods and services continue to be delivered in an economically viable way, while at the same time minimising the impact on the ecosystem services and maximising social benefits (see Figure 2). Only then can we hope to achieve the development that 'meets the needs of the present without compromising the ability of future generations to meet their needs'.[3]

This is by no means a trivial task as it requires action by all actors in society, including governments, industry and individuals. One of the challenges is that, even if there was a universal commitment to sustainable development, it is still unclear what goods, services and activities are sustainable and how they could be identified.

In an attempt to contribute towards a better understanding of this complex problem, this chapter concentrates on one dimension of sustainable development – the environment – to illustrate how environmentally sustainable products and activities can be identified for a more sustainable management of ecosystem services. It is argued that this must be underpinned by life cycle thinking in order to understand fully the interactions between industrial systems, human activities and ecosystem services.

After a brief overview of life cycle thinking and the Life Cycle Assessment (LCA) methodology, the rest of the chapter presents a number of case studies to demonstrate how LCA can be used as a tool for a more sustainable management of ecosystem services.

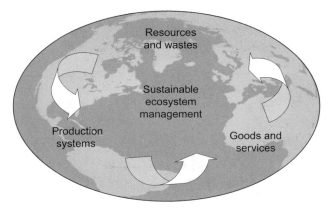

Figure 2 Moving from a 'three- to one-planet economy': a more sustainable approach to managing ecosystem services must be found

2 Life Cycle Thinking and Life Cycle Assessment

Estimations of environmental impacts have traditionally focused on one life cycle stage, most often manufacturing. This has led to the development of end-of-pipe legislation and various clean-up technologies. However, while this approach reduces the immediate pollution from industrial installations, the use of energy and chemicals and the need to further treat and dispose of the wastes generated in the clean-up process often lead to additional pollution further up- or downstream of that industrial facility. Thus, instead of protecting the environment, considering one stage in the life cycle can lead to higher overall environmental impacts. Similarly, depending on the product, manufacturing may not be environmentally the most important stage in the life cycle and the product use or post-consumer waste management may be a 'hot spot' instead. Therefore, we can only be certain that we are protecting the environment as a whole if we adopt a systems approach to consider the whole life cycle of an activity. This is known as 'life cycle thinking' or a 'life cycle approach'. As illustrated in Figure 3, taking a life cycle approach means drawing the system boundary from 'cradle to grave' – or in other words, considering the whole life cycle of a product, process or activity from the extraction of raw materials, to production of the product or provision of a service to the use and end-of-life stages.

It is now widely accepted that environmentally sustainable options can only be found by taking a life cycle approach to managing ecosystem services. In this way we can obtain a full picture of human interactions with the environment and avoid shifting of environmental impacts from one life cycle stage to another.

Life Cycle Assessment (LCA) is an environmental management tool that helps to translate qualitative life cycle thinking into a quantitative measure of environmental sustainability of products, processes or activities on a life cycle

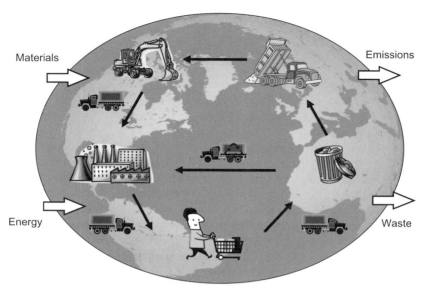

Figure 3 A life cycle approach to understanding and managing ecosystem services from 'cradle to grave'.

basis. By taking into account the whole life cycle of a product or activity along the supply chain, LCA enables identification of the most significant impacts and stages in the life cycle that should be targeted for maximum improvements. The following section gives a brief overview of the LCA methodology.

2.1 LCA Methodology: An Overview

Although LCA has been used in some industrial sectors, particularly energy, for over 20 years, it has only received wider attention and methodological developments since the beginning of the 1990s, when its relevance as an environmental management aid in both corporate and public decision-making became more evident.[4] Some examples of this include incorporation of life cycle thinking and LCA within the ISO 14000 Environmental Management Systems (EMS)[5] and the EC Directive on Integrated Pollution Prevention and Control (IPPC),[6] which require companies to have a full knowledge of the environmental consequences of their actions, both on and off site.

Today, LCA is a well-established tool and is used in a variety of applications in industry, research and policy making. Some of the applications include identification of environmental sustainability indicators; measuring the environmental sustainability of products and technologies; identification of the most dominant stages or 'hot spots' in the life cycle of products and processes; identification of improvement options; product design; and process optimisation. A review and examples of some of these applications of LCA to various products and processes can be found, for example, in work by Azapagic.[7,8]

The LCA methodology is standardised by the ISO 14040[9] and 14044[10] standards. As defined by ISO 14040, LCA is a compilation and evaluation of the inputs, outputs and the potential environmental impacts of a product throughout its life cycle, from acquisition of raw materials through production, use and waste disposal (*i.e.* from cradle to grave). Figure 3 shows the life cycle stages normally considered in an LCA of a product. Although the ISO standards refer to products only, LCA can also be used to calculate the environmental impacts of processes, technologies,[7,8] services or activities.[10]

The LCA methodology comprises the following four phases:[10]

1. Goal and scope definition;
2. Inventory analysis;
3. Impact assessment; and
4. Interpretation.

A brief overview of each phase is given below.

2.1.1 Goal and Scope Definition

The process of conducting an LCA, as well as its outcomes, is largely determined by the goal and scope of the study. For example, the goal of the study may be to identify the 'hot spots' in a manufacturing process and to use the results internally by a company to reduce the environmental impacts from the process. Alternatively, the company may wish to use the result externally, either to provide the LCA data to customers who use their product as a raw material, or perhaps to market their product on the basis of the LCA results. In each case, the assumptions, data and system boundaries may be different so that it is important that these are defined in accordance with the goal of the study.

In full LCA studies the system boundary is drawn to encompass all stages in the life cycle from extraction of raw materials to the final disposal, *i.e.* from 'cradle to grave' (see Figure 3). However, in some cases, the scope of the study will demand a different approach, where it is not appropriate or even possible to include all stages in the life cycle. This is usually the case, for example, with chemical commodities and intermediate products, which can have a number of different uses so that it is not possible to follow their numerous life cycles after the production stage. The scope of such studies can be from 'cradle to gate' as they follow a product from the extraction of raw materials to the factory gate.

One of the most important steps in LCA is the definition of the functional unit. The functional unit represents a quantitative measure of the output(s) that the system delivers. In comparative LCA studies it is crucial that alternative systems are compared on the basis of an equivalent function, *i.e.* the functional unit. For example, comparison of different beverage packaging should be based on their equivalent function, which is to contain a certain amount of beverage. The functional unit is then defined as 'the quantity of packaging necessary to contain the specified volume of beverage'.[4]

This phase also includes assessment of data quality with respect to time, geographical location and technologies covered in the study. Assumptions and limitations of the study should also be stated clearly in this phase.

Goal and scope are constantly reviewed and refined during the process of carrying out an LCA, as additional data and information become available.

2.1.2 Inventory Analysis

Life Cycle Inventory (LCI) analysis involves collection of environmental burdens data necessary to meet the goals of the study. The environmental burdens (or interventions) comprise the materials and energy used in the system, emissions to air, liquid effluents and solid wastes discharged into the environment.

Following a preliminary system definition in the goal and scope definition phase, detailed system specification is carried out in the LCI phase to identify data needs. A system is defined as a collection of materially and energetically connected operations (for example, a manufacturing process) which performs some defined function. Detailed system characterisation involves its disaggregation into a number of inter-linked subsystems. Environmental burdens are then quantified for each subsystem according to Equation (1):

$$B_j = \sum_{i=1}^{I} b_{j,i} x_i \qquad (1)$$

where $b_{j,i}$ is burden (or intervention) j from process or subsystem i and x_i is a mass or energy flow associated with that subsystem.

If the system under study produces more than one functional output, then the environmental burdens from the system must be allocated among these outputs. This is the case, for example, with co-product, re-use and recycling systems; such systems in LCA are known as 'multiple-function systems'. Allocation is the process of assigning to each function of a multiple-function system only those environmental burdens that that functional output is responsible for. In ISO 14044[10] three methods are recommended for dealing with allocation:

1. If possible, allocation should be avoided by disaggregating the given process into different sub-processes or by system expansion.
2. If it is not possible to avoid allocation, then the allocation problem must be solved by using system modelling which reflects the underlying physical relationships among the functional units.
3. Where physical relationships cannot be established, other relationships, including economic value of the functional outputs, can be used.

The allocation method used will usually influence the results of the LCA study so the identification of an appropriate allocation method is crucial. Sensitivity analysis should be carried out in cases where the use of different allocation methods is possible to determine the influence of the allocation method on the results.

2.1.3 Impact Assessment

Life Cycle Impact Assessment (LCIA) is the third LCA phase and its main purpose is to translate the environmental burdens quantified in LCI into the related potential environmental impacts (or category indicators). This is carried out within the following three steps:

1. Selection of impact categories, category indicators and LCIA models;
2. Classification;
3. Characterisation.

The selection of impact categories, category indicators and LCIA models must be consistent with the goal and scope of the LCA study and must reflect the environmental issues of the system under study. Classification involves aggregation of environmental burdens into a smaller number of environmental impact categories to indicate their impacts on human and ecological health and the extent of resource depletion. The identification of impacts of interest is then followed by their quantification in the next, characterisation step, as follows:

$$E_k = \sum_{j=1}^{J} ec_{k,j} B_j \qquad (2)$$

where $ec_{k,j}$ represents characterisation factor k for burden B_j showing its relative contribution to impact E_k. The characterisation factors are calculated using appropriate LCIA models.

A number of LCIA methods exist, and they are divided into two general groups:

1. Problem-oriented approaches;
2. Damage-oriented methods.

In the problem-oriented methods the environmental burdens are aggregated according to their relative contribution to the environmental effects that they might cause. They are often referred to as 'midpoint' approaches because they link the environmental interventions from LCI somewhere in between the point of intervention and the ultimate damage caused by that intervention (see Figure 4). Damage-oriented methods, on the other hand, model the 'endpoint' damage caused by environmental interventions to 'areas of protection', which include human health, natural and human-made environment.[11] The most widely used problem-oriented method is the CML 2 method;[12] Eco-Indicator 99[13] is the most commonly used damage-oriented method. Overall, the CML 2 method is more prevalent in LCA studies, so a brief overview of this approach is provided in the following paragraph.

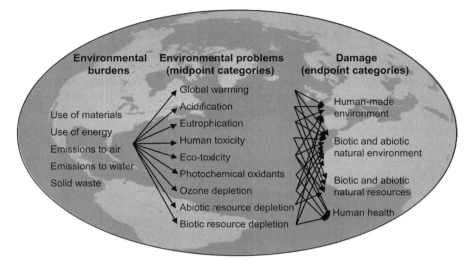

Figure 4   Link between environmental interventions, problems (midpoint categories) and damage (endpoint categories) to the environment and human health.

In the CML 2 method, environmental burdens are aggregated according to their relative contributions to the environmental problem or impact that they can potentially cause. The following impacts, defined as midpoint categories, are considered in this method (see Figure 4):

- Abiotic resource depletion;
- Global warming;
- Ozone depletion;
- Acidification;
- Eutrophication;
- Photochemical oxidant formation (photochemical or summer smog);
- Human toxicity; and
- Ecotoxicity (freshwater, marine and terrestrial).

Their definitions can be found in Appendix 1. As shown in the Appendix, the impacts are calculated relative to the characterisation factor of a reference substance. For example, carbon dioxide (CO_2) is a reference gas for determining global warming potentials of other related gases, such as methane (CH_4) and nitrous oxide (N_2O); therefore, its characterisation factor is 1 kg CO_2 eq/kg CO_2 whilst that of CH_4 is 25 kg CO_2 eq/kg CH_4. Note that the equations for calculating the impacts given in the Appendix are based on the general Equation (2) for characterisation of environmental impacts, defined in section 2.1.3.

The impacts calculated by this method are categorised as potential, rather than actual, as they are quantified at the intermediate position between the point of environmental intervention and the damage caused, rather than at the endpoint.

Returning to the remaining stages within the LCIA phase, three further optional steps are:

1. Normalisation;
2. Grouping;
3. Weighting of impacts.

The impacts can be normalised with respect to the total emissions or extractions in a certain area and over a given period of time. This can help to assess the extent to which an activity contributes to the regional or global environmental impacts. However, normalisation results should be interpreted with care because of the lack of reliable data for many impacts on both the regional and global scales.

Grouping involves qualitative or semi-quantitative sorting and/or ranking of impacts and it may result in a broad ranking or hierarchy of impact categories with respect to their importance. For example, categories could be grouped in terms of high importance, moderate importance and low priority issues.

The final optional stage within LCIA is weighting of impacts, often referred to as valuation. It involves assigning weights of importance to the impacts to indicate their relative importance. As a result, all impact categories are aggregated into a single environmental impact function *EI* as follows:

$$EI = \sum_{k=1}^{K} w_k E_k \qquad (3)$$

where w_k is the relative importance of impact E_k.

Weighting is the most subjective element of LCA because it involves social, political and ethical value choices. At present, there is no consensus on how to aggregate the environmental impacts into a single environmental impact function or even on whether such aggregation is conceptually and philosophically valid.

2.1.4 Interpretation

The main objectives of this phase are to analyse results, reach conclusions, explain limitations and provide recommendations based on the findings of LCI and/or LCIA. Quantification of environmental impacts carried out in LCI and LCIA enables identification of the most significant issues and life cycle stages that contribute to these issues. This information can then be used to target these 'hot spots' for system improvements or innovation.

Before the final conclusions and recommendations of the study are made it is important to carry out sensitivity analysis. Data availability and reliability are some of the main issues in LCA, since the results and conclusions of an LCA study will be determined by the data used. Sensitivity analysis can help identify the effects that data variability, uncertainties and data gaps have on the final results of the study and indicate the level of reliability of the final results of the study.

Finally, the findings and conclusions of the study are reported in accordance to the intended use of the study. The report should give a complete, transparent and unbiased account of the study as detailed in ISO 14044.[10] If the study is used externally, critical review by an independent agent should be carried out.

Further detail on the LCA methodology can be found in the ISO 14040 and 14044 standards,[9,10] respectively.

3   LCA as a Tool for Sustainable Management of Ecosystem Services

Due to its ability to quantify environmental interventions and the related impacts, LCA lends itself naturally as a tool for assessing and managing the environmental sustainability of ecosystems, particularly provisioning services. The use of LCA for these purposes can be to:

- Quantify environmental emissions and impacts;
- Identify hot spots in the system;
- Identify opportunities for improvements; and
- Enable comparison of alternative products or services.

In addition, LCA can help to identify impacts on supporting ecosystem services, for example, through quantifying nutrient and water requirements and resource depletion, as well as informing regulation services, *e.g.* on the needs for air quality and water regulation.

In this chapter, the focus is on the provisioning services and the discussion that follows illustrates how LCA can be used to quantify environmental impacts of different products and services and to identify more sustainable alternatives. Four major human activities and supply chains, which contribute most to the global environmental impacts,[14] are chosen for discussion here:

- Energy;
- Transport;
- Industry; and
- Agriculture and food.

The discussion centres on Global Warming Potential (or carbon footprint), as one of the major global impacts; however, other impacts are also discussed as appropriate. Note that all the impacts have been estimated using the CML 2 method (see Appendix 1).

3.1   Life Cycle Impacts of Energy: The Electricity Sector

In 2004, the global direct emissions of greenhouse gases were estimated at 49 Gt of CO_2 eq,[14] of which energy generation and supply contributed 14.72 Gt or

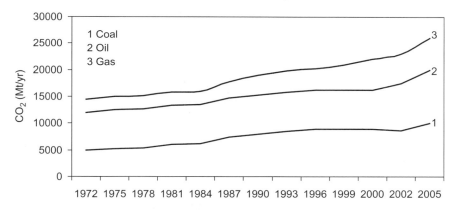

Figure 5 Global direct emissions of CO_2 by fuel.[14]

26%. As shown in Figure 5, fossil fuel energy contributes to the vast majority of direct CO_2 emissions. Estimates of the global emissions from energy supply on a life cycle basis are not available, but it is clear that the life cycle emissions would be significantly higher than the direct emissions. As an example of the life cycle impacts of energy supply, electricity generation using different technologies is considered here. The life cycle of electricity supply can be seen in Figure 6. As shown, the life cycle involves extraction of fuels and construction materials; conversion of fuels or energy carriers to electricity (depending on the technology used); distribution of electricity; and its use. Each stage in the life cycle uses natural resources (materials and energy) and discharges wastes (gaseous, liquid and solid) into the environment.

Figure 7 compares the global warming potential (carbon footprint) of different electricity-generating options. Of the options compared, electricity from lignite has the highest carbon footprint, equal to $1220 \text{ g } CO_2$ eq/kWh; heavy fuel oil and hard coal follow closely with 1130 and $1100 \text{ g } CO_2$ eq/kWh, respectively, while electricity from a gas plant of the comparable size generates less than half that amount ($430 \text{ g } CO_2$ eq/kWh). The most sustainable options from the carbon point of view are wind, hydropower, solar thermal and nuclear plants. Therefore, if we were only concerned with the carbon equivalent emissions from the life cycle of electricity generation, these four options would help us to become more sustainable. However, if compared for a range of other life cycle impacts, the order of preference for different technologies changes. This is shown in Figure 8.

Two things can be observed in Figure 8. First, the heavy fuel oil option has significantly higher impacts than lignite for all impacts (up to 4.5 times), except for freshwater toxicity (which is 2.7 times lower than for lignite). Therefore, if a wider range of impacts is considered, rather than just global warming potential, then between heavy fuel oil and lignite, the latter could arguably be selected as a more sustainable option overall. Obviously, this would depend on the relative

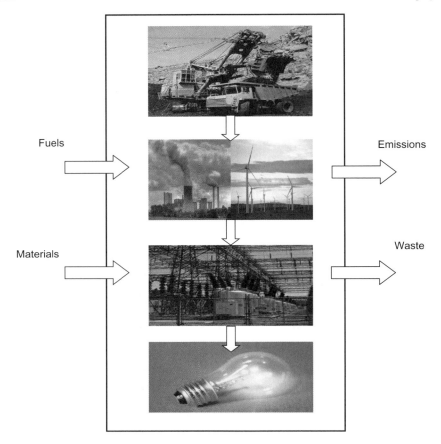

Figure 6   The life cycle of electricity generation (for illustration purposes, only fossil-
fuel and wind options are shown).

importance of different impacts, however, as well as on other sustainability
aspects, such as the long-term availability of oil *versus* coal and the respective
costs of the fuels. Ultimately, neither option is sustainable in the medium to
long term and other options need to be considered, particularly renewables and
nuclear.

This brings us to the second point to observe from Figure 8: for several
impacts, some of the renewable options come close to or exceed the impacts
from some of the fossil fuel options, notably gas. For example, both the solar
photovoltaic (PV) and hydro (pumped storage) options have higher acidification impact than the best gas option (combined cycle): by a factor of 1.5 and
1.2 times, respectively. The PV option also has a much higher eutrophication
impact than any of the fossil fuel options, except for heavy fuel oil: 2 times
higher than the lignite and 4.7 times higher than gas (100 MW, steam turbine).
Solar PV, wind and hydro also exceed toxicity of some of the fossil fuel options

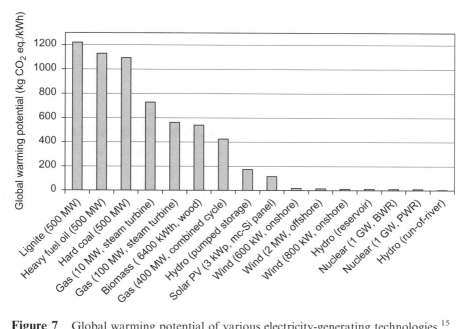

Figure 7 Global warming potential of various electricity-generating technologies.[15]

(*e.g.* gas and hard coal). These impacts occur at different stages in the life cycle, including the raw materials and manufacturing.

Thus, this simple example demonstrates the complexity of identifying more sustainable energy options, even before other sustainability issues, such as costs and social acceptability, are considered. LCA can help towards identifying the most significant impacts and possibly more environmentally-sustainable options; however, the ultimate choice of sustainable options will depend on the decision-making context and decision-makers' preferences.

3.2 Life Cycle Impacts of Transport

Transport was responsible for 13% or 6.4 Gt of the direct global CO_2 eq emissions in 2004.[14] The majority of these emissions are from road transport (75%) and aviation (12%). Similar to the energy sector, the global life cycle impacts of transport are not available, so in this section we consider a couple of examples, related to different transportation modes and fuels.

As shown in Figure 9, the life cycle of the transport sector involves extraction of fuels and raw materials; manufacture of transportation vehicles; transportation and end-of-life vehicle waste management. Each stage involves the use of natural resources and emissions to the environment.

In Figure 10 different transportation modes are compared according to their carbon footprint. The results suggest that travelling by train is the best option with 8 and 20 g CO_2 eq per person and kilometre travelled for long-distance and

Adisa Azapagic

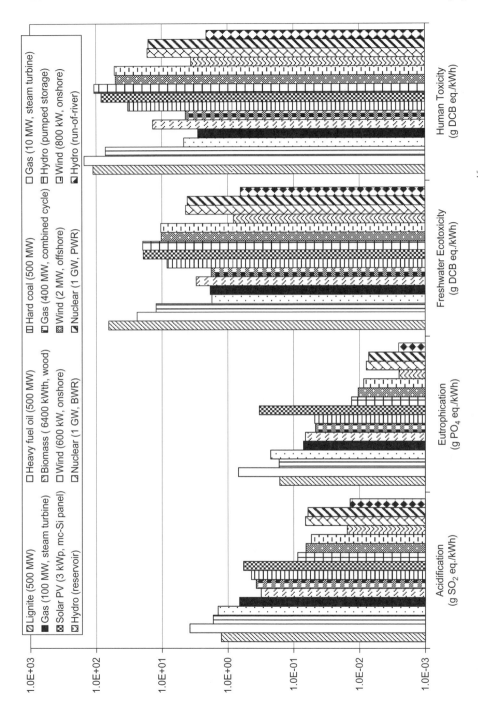

Figure 8 Comparison of electricity-generating technologies for different life cycle environmental impacts.[15]

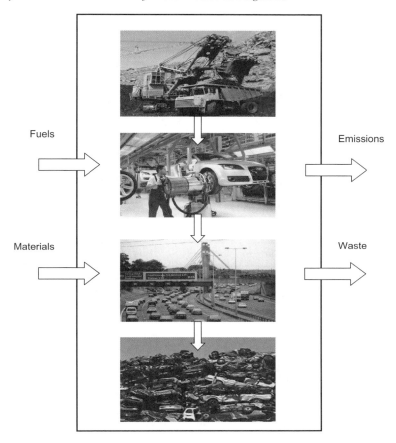

Figure 9   The life cycle of transport (for illustration purposes, only road transport shown).

regional travel, respectively. Short-haul flights are the worst option with 330 g CO_2 eq/person km; travel by car generates just over half that amount (180 g CO_2 eq/person km). Interestingly, travel by bus and long-distance flying have a comparable carbon footprint per distance travelled (125 CO_2 eq/person km). Therefore, these results give an indication as to which transport options are more sustainable – however, as previously, other aspects would need to be taken into account, such as travel costs and availability of different transport options in different regions.

It is also interesting to look at the life cycle carbon footprints of different biofuels, which have been hailed as a sustainable replacement for the fossil-based fuels. As shown in Figure 11, bioethanol from the UK wheat offers a saving of about 28% of the carbon footprint compared to that of petrol. However, this comparison assumes a 100% replacement of petrol by the

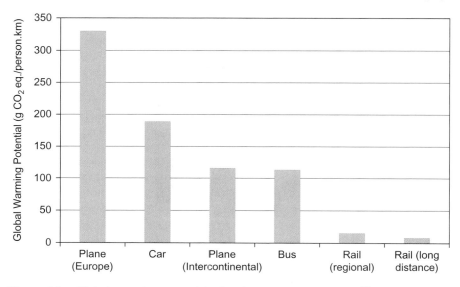

Figure 10   Global warming potential of various transport options.[15]

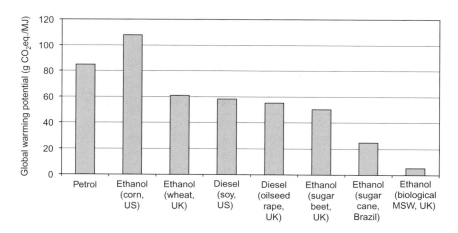

Figure 11   Global warming potential of different biofuels in comparison with petrol.[16,17]

biofuel – adding only 10% to petrol, as is currently the case, results in marginal savings of carbon. The best performing bioethanol is from the Brazilian sugar cane, saving 70% of CO_2 eq compared to petrol, while the bioethanol from US corn has a 27% higher carbon footprint than that of petrol (both at a 100% replacement). In addition to the other issues such as competition with food production, these results further confirm that the first-generation biofuels are unsustainable and that we need to turn our attention towards the second-generation of biofuels.

3.3 Life Cycle Impacts of Industry: The Example of the Chemical Sector

Industrial activities were directly responsible for 14% or 6.86 Gt of CO_2 eq in 2004.[14] Two thirds of these emissions came from three sectors: iron and steel, non-metallic minerals and petrochemicals. The latter is considered in this section as an example.

The life cycle of the chemicals supply chain is shown in Figure 12, involving extraction and refining of raw materials and fuels, manufacture and use of chemicals and end-of-life management.

In 2005, the life cycle CO_2 eq emissions from the chemical industry amounted to 3.3 Gt CO_2 eq $(+/-25\%)$.[18] As can be seen from Figure 13, 63% of the emissions or 2.1 Gt CO_2 eq are a result of the production of chemicals. This figure includes direct and indirect energy use in the production, as well as process emissions; 0.3 Gt CO_2 eq arise during the extraction phase of the

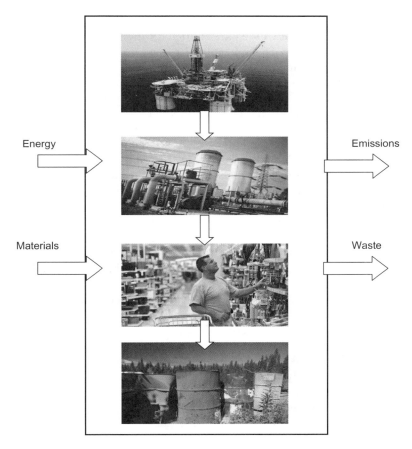

Figure 12 The life cycle of chemicals.

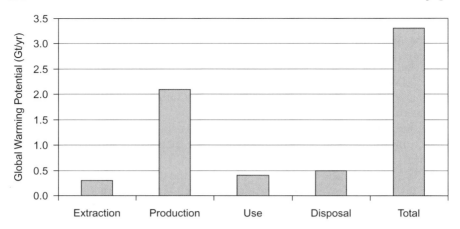

Figure 13   Contribution to global warming potential of different stages in the life cycle
of chemicals (based on global production in 2005).[18]

feedstock and fuels; $0.5\,Gt\ CO_2$ eq are emitted during the disposal phase of the
produced chemicals and $0.4\,Gt$ are emitted by end users.

It is interesting to compare the carbon footprints of different chemicals. As
an example, a study was carried out of chemical commodities produced in the
UK, with the aim of estimating the life cycle carbon footprints of different
chemicals based on their annual production volumes.[19] These results are given
in Figure 14 and Figure 15, with the former showing the carbon footprints per
tonne of each chemical and the latter the total annual carbon footprint.

As shown in Figure 14, with $8720\,kg\ CO_2$ eq/tonne, ammonium nitrate has
the highest carbon footprint of all chemicals considered here; the carbon
footprint of the second-worst chemical, aniline, is around half that amount
($4950\,kg\ CO_2$ eq/tonne). At around $1400\,kg\ CO_2$ eq, ethylene and hydrochloric
acid have the lowest carbon footprint per tonne of product. However, when
compared on the basis of their annual production, the ranking changes (see
Figure 15). Although ammonium nitrate still has the highest total carbon
footprint with 4.2 Mt of CO_2 eq/yr, ethylene now has the second highest
impact, equal to 2.2 Mt of CO_2 eq/yr. Polypropylene follows with 1.6 Mt of
CO_2 eq/yr, and then benzene and low-density polyethylene (LDPE) with 1.3 Mt
of CO_2 eq/yr each. This information can be useful at a sectoral and a national
level, as it can help inform relevant decision makers as to which products can
help the UK achieve the Kyoto targets. However, care should also be taken in
interpreting these results, as these chemicals serve different purposes and cannot be compared directly.

It is interesting to note in the context of this analysis that ammonium nitrate,
the fertiliser which helped current societies to develop through improved food
production, is now deemed less sustainable due to its contribution to global
warming. Many other products may follow the same fate, in the same way that,
for example, the CFCs did. On this note, paradoxically, the HCFCs were phased

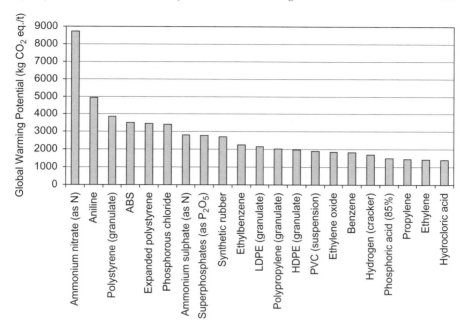

Figure 14 Global warming potential of different chemicals (system boundary from 'cradle to gate').

in to replace the ozone-depleting CFCs, only to cause another environmental problem – global warming – and are now themselves subject to replacement.

As has been mentioned previously, a comparison based on only one environmental impact should be carried out with caution, otherwise unsustainable decisions could be made. As an illustration, Figure 16 compares several chemicals on other environmental impacts, including the chemicals with the highest and lowest carbon footprints. Ammonium sulfate is also included in the comparison, as a possible alternative to ammonium nitrate as a fertiliser. The results show that ammonium nitrate remains the least sustainable option for all the impacts considered and ethylene has the lowest impacts overall. The impact of the hydrochloric acid (HCl) significantly exceeds that of ethylene and in the case of toxicity, by up to 770 times. Similarly, marine toxicity of HCl exceeds that of ammonium sulfate by 15%. Thus, this simple example demonstrates once again the importance of considering wider environmental (and other sustainability) issues, since focusing on one aspect of (environmental) sustainability can lead to sub-optimal decisions.

3.4 Life Cycle Impacts in the Food Sector

As shown in Figure 17, the life cycle of food provision includes crop cultivation and animal rearing; food production and preparation; and waste management.

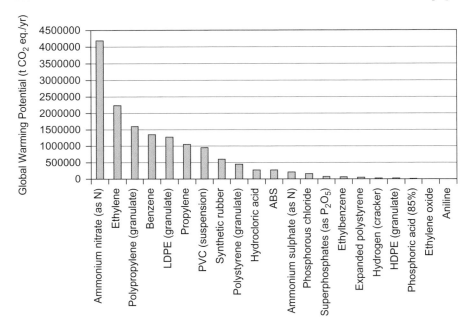

Figure 15 Total global warming potential of different chemicals produced in the UK (based on average production data for 2002–2006; system boundary from 'cradle to gate').

The global life cycle impacts of food are unknown, but it is estimated that direct greenhouse emissions from agriculture contribute 13.5% or 6.6 Gt CO_2 eq of global emissions, with the main contributors being fertilisers (38%), livestock (31%) and rice production (11%).[14]

For illustration of the life cycle impacts of food systems, the carbon footprint of several types of vegetables and meat are given in Figure 18. As can be seen, potato has a low carbon footprint compared to tomato (the latter is grown in a greenhouse heated by electricity). In fact, the tomato from a greenhouse has a higher carbon footprint per kilogram than pork and turkey. Beef and lamb have the highest carbon footprint, equal to 14 kg CO_2 eq per kg of meat. Therefore, this would suggest that a vegetarian diet is environmentally more sustainable than eating meat; however, other factors, such as personal preferences, will influence the choice of diet.

Currently, there is a lively activity on providing information to consumers on the life cycle impacts of various food and other products so that environmentally more informed purchasing decisions can be made. This is being effected in the UK through the carbon label, developed by the BSI.[21] Orange juice is an example of a product that has been carbon-labelled; the carbon footprint results are shown in Figure 19. Overall, it would appear that orange juice from concentrate, kept at ambient conditions, is the best option. It is interesting to note that freshly squeezed orange juice has a higher environmental impact than the

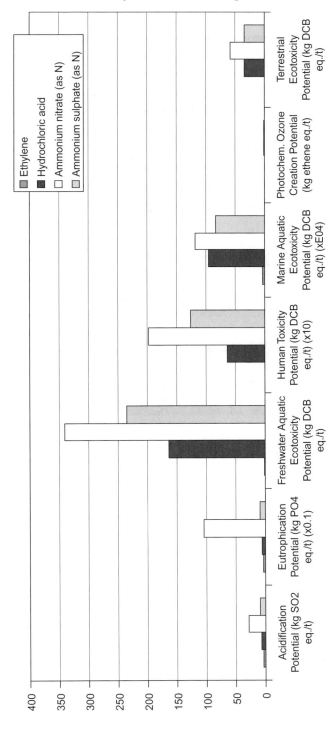

Figure 16 Environmental impacts of chemicals: selected examples.

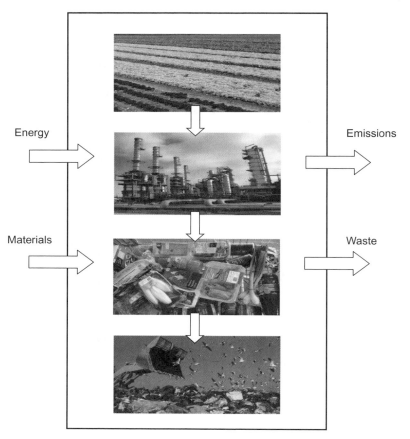

Figure 17   The life cycle of food production.

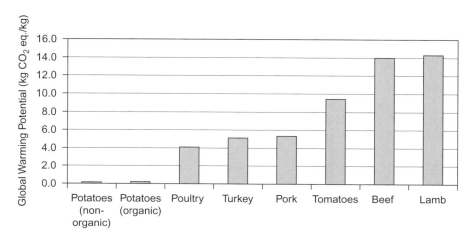

Figure 18   Global warming potential of different foodstuffs from 'cradle to farm gate'.[20]

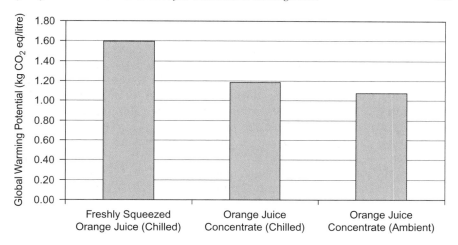

Figure 19 The carbon footprint of orange juice produced in Brazil.[22]

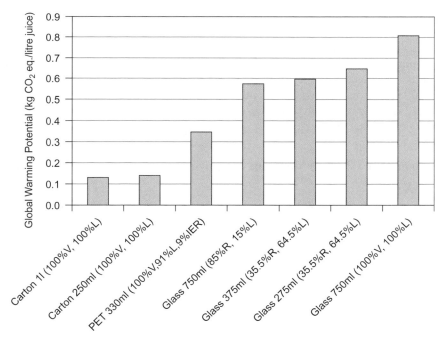

Figure 20 The carbon footprint of different juice packaging.[23]

juice from concentrate – this is due to a higher impact of transport compared with that of dewatering the juice (to make the concentrate), as the product is imported from Brazil. This is one of the rare examples where transport (known as 'food miles') plays an important role – in most food production systems, the impact of transport is negligible and the agricultural production is responsible for a large proportion of the impact.

Packaging is another point of debate in terms of environmental impacts. Again as an illustration, Figure 20 shows the carbon footprint for selected orange juice packaging. Overall, the carton has the lowest impact and glass the highest. Compared to the carbon footprint of orange juice (ambient concentrate), carton contributes around 10% to the total impact, while glass packaging would almost double the impact of the juice. Therefore, the contribution of packaging depends on the type of product and the type of packaging – for carbon-intensive products (*e.g.* meat), packaging has little impact, while for other products (*e.g.* non-alcoholic drinks) the contribution of packaging to the carbon footprint can range widely (20–90%).

Conclusions

This paper has discussed and illustrated how LCA can be used as a tool for more sustainable management of ecosystem services. Four industrial sectors and supply chains have been discussed: energy, transport, industry and food. The results presented demonstrate how this information could be used to compare and identify more sustainable options.

Much more effort is needed in measuring sustainability of a wide range of products and human activities to help industry, consumers and policy makers identify more sustainable options. It should also be borne in mind that sustainable solutions can only be identified by considering a range of sustainability issues, as focusing on single issues – such as carbon footprints – may lead to an increase in other impacts.

Appendix 1 CML 2 Method: Definition of Environmental Impact Categories

Global Warming Potential (GWP) is calculated as the sum of emissions of greenhouse gases multiplied by their respective GWP factors, GWP_j:

$$GWP = \sum_{j=1}^{J} GWP_j B_j \qquad \text{kg } CO_2 \text{ eq}$$

where B_j represents the emission of greenhouse gas j. GWP factors for different greenhouse gases are expressed relative to the global warming potential of CO_2, which is therefore unity. The values of GWP depend on the time horizon over which the global warming effect is assessed. GWP factors for shorter times (20 and 50 years) provide an indication of the short-term effects of greenhouse gases on the climate, while GWP for longer periods (100 and 500 years) are used to predict the cumulative effects of these gases on the global climate.

Acidification Potential (AP) is based on the contribution of sulfur dioxide (SO_2), nitrogen oxides (NOx) and ammonia (NH_3) to the potential acid deposition. AP is calculated according to the equation:

$$AP = \sum_{j=1}^{J} AP_j B_j \qquad \text{kg } SO_2 \text{ eq}$$

where AP_j represents the acidification potential of gas j expressed relative to the AP of SO_2 and B_j is its emission in kg.

Eutrophication Potential (EP) is defined as the potential of nutrients to cause over-fertilisation of water and soil, which can result in increased growth of biomass. It is calculated as:

$$EP = \sum_{j=1}^{J} EP_j B_j \qquad \text{kg } PO_4^{3-} \text{ eq}$$

where B_j is an emission of species such as nitrogen (N), nitrogen oxides (NOx), ammonium (NH_4^+), phosphate (PO_4^{3-}), phosphorus (P), and chemical oxygen demand (COD); EP_j represents their respective eutrophication potentials. EP is expressed relative to PO_4^{3-}.

Human Toxicity Potential (HTP) is calculated by taking into account releases toxic to humans to three different media, *i.e.* air, water and soil:

$$HTP = \sum_{j=1}^{J} HTP_{jA} B_{jA} + \sum_{j=1}^{J} HTP_{jW} B_{jW} + \sum_{j=1}^{J} HTP_{jS} B_{jS} \text{ kg } 1,4\text{-DB eq}$$

where HTP_{jA}, HTP_{jW}, and HTP_{jS} are toxicological classification factors for substances emitted to air, water and soil, respectively, and B_{jA}, B_{jW} and B_{jS} represent the respective emissions of different toxic substances into the three environmental media. The reference substance for this impact category is 1,4-dichlorobenzene.

Eco-Toxicity Potential (ETP) is also calculated for all three environmental media and comprises five indicators ETP_n:

$$ETP_n = \sum_{j=1}^{J} \sum_{i=1}^{I} ETP_{i,j} B_{i,j} \qquad \text{kg } 1,4\text{-DB eq}$$

where n ($n = 1$–5) represents freshwater and marine aquatic toxicity; freshwater and marine sediment toxicity and terrestrial ecotoxicity, respectively. $ETP_{i,j}$ represents the ecotoxicity classification factor for toxic substance j in the compartment i (air, water and soil) and $B_{i,j}$ is the emission of substance j to compartment i. ETP is based on the maximum tolerable concentrations of different toxic substances in the environment by different organisms. The reference substance for this impact category is also 1,4-dichlorobenzene.

Photochemical Oxidants Creation Potential (POCP) is related to the potential of volatile organic compounds (VOCs) and NOx to generate photochemical or summer smog. It is usually expressed relative to the POCP classification factor of ethylene and can calculated as:

$$POCP = \sum_{j=1}^{J} POCP_j B_j \qquad \text{kg ethylene eq}$$

where B_j is the emission of species j participating in the formation of summer smog and $POCP_j$ is its classification factor for photochemical oxidation formation.

Abiotic Resource Depletion Potential (ADP) includes depletion of fossil fuels, metals and minerals. The total impact is calculated as:

$$ADP = \sum_{j=1}^{J} ADP_j B_j \qquad \text{kg Sb eq}$$

where B_j is the quantity of abiotic resource j used and ADP_j represents the abiotic depletion potential of that resource. This impact category is expressed in kg of antimony used, which is taken as the reference substance for this impact category. Alternatively, kg oil eq can be used instead.

Stratospheric Ozone Depletion Potential (ODP) indicates the potential of emissions of chlorofluorohydrocarbons (CFCs) and other halogenated hydrocarbons to deplete the ozone layer and is expressed as:

$$ODP = \sum_{j=1}^{J} ODP_j B_j \qquad \text{kg CFC} - 11 \text{ eq}$$

where B_j is the emission of ozone depleting gas j. The ODP factors are expressed relative to the ozone depletion potential of CFC-11.

For a full description of the CML 2 methodology, see Guinée *et al.*[12]

References

1. WWF, Zoological Society of London and Global Footprint Network, *Living Planet Report* 2008, WWF, Zoological Society of London and Global Footprint Network, UK, 2008. http://da.cop15.dk/files/pdf/LPR_2008.pdf.
2. WRI, Millennium Ecosystem Assessment, 2005, *Ecosystems and Human Well-being: Biodiversity Synthesis*, World Resources Institute, Washington, DC, USA, 2005.
3. The Brundtland Commission, *Our Common Future, The Report of the World Commission on Environment and Development*, Oxford University Press, Oxford, UK,1987.

4. A. Azapagic, in *Renewables-based Technology: Sustainability Assessment*, ed. J. Dewulf and H. van Langenhove, John Wiley & Sons, London, 2006, ch. 6, pp. 87–110.
5. ISO, *ISO 14001: Environmental Management Systems – Specification with Guidance for Use*, ISO, Geneva, Switzerland, 1996.
6. *EC Council Directive 91/61/EC: Concerning Integrated Pollution Prevention and Control*, Off. J. Eur. Comm., HMSO, London, No. L257, 1996.
7. A. Azapagic, *Chem. Eng. J.*, 1999, **73**, 1–21.
8. A. Azapagic, in *Handbook of Green Chemistry and Technology*, ed. J. Clark and D. Macquarrie, Blackwell Science, Oxford, 2002, pp. 62–85.
9. ISO, *ISO/DIS 14040: Environmental Management – Life Cycle Assessment – Principles and Framework*, ISO, Geneva, Switzerland, 1997.
10. ISO, *ISO/DIS 14044: Environmental Management — Life Cycle Assessment — Requirements and Guidelines*, ISO, Geneva, Switzerland, 2006.
11. H. A. Udo de Haes and E. Lindeijer, in *Life-cycle Impact Assessment: Striving Towards Best Practice,* ed. H. Udo de Haes, O. Jolliet, G. Finnveden, M. Goedkoop, M. Hauschild and E. Hertwich, SETAC Press Pensacola, FL, USA, 2002.
12. J. B. Guinée, M. Gorrée, R. Heijungs, G. Huppes, R. Kleijn, L. van Oers, A. Wegener Sleeswijk, S. Suh, H. A. Udo de Haes, H. de Bruijn, R. van Duin and M. A. J. Huijbregts, *Life Cycle Assessment: An Operational Guide to the ISO Standards. Parts 1, 2a & 2b,* Kluwer Academic Publishers, Dordrecht, The Netherlands, 2001.
13. M. Goedkoop and R. Spriensma, *The Eco-Indicator 99: A Damage Oriented Method for Life Cycle Assessment, Methodology Report*, 22 June 2001, Pré Consultants, Amersfoort, The Netherlands, 3rd edn, 2001.
14. *IPCC, Fourth Assessment Report: Climate Change 2007.*
15. Ecoinvent Database, Swiss Centre for Life Cycle Inventories. http://www.ecoinvent.ch/.
16. Department for Transport, *Carbon and Sustainability Reporting within the Renewable Transport Fuel Obligation*, Government Recommendation to RTFO Administrator, Department of Transport, UK, June 2007.
17. H. Stichnothe and A. Azapagic, *Resour. Conserv. Recycl.*, 2009, **53**, 624–630.
18. International Council of Chemical Associations and Responsible Care, Innovations for Greenhouse Gas Reductions: A Life Cycle Quantification of Carbon Abatement Solutions Enabled by the Chemical Industry, July 2009. http://www.icca-chem.org/ICCADocs/LCA-executive-summary-english1.pdf.
19. D. Fairweather and A. Azapagic, *The Carbon Footprint of the UK Chemical Industry,* The University of Manchester, Manchester, 2009.
20. A. G. Williams, E. Audsley and D. L. Sandars, *Final Report to Defra on Project IS0205: Determining the Environmental Burdens and Resource Use in the Production of Agricultural and Horticultural Commodities,* Defra, London, 2006.

21. BSI, PAS2050: 2008. *Specification for the Assessment of The Life Cycle Greenhouse Gas Emissions of Goods and Services*, BSI, October 2008.
22. Tesco, The Carbon Footprint of Orange Juice, Tesco, 2008. http://www.tesco.com/greenerliving/greener_tesco/what_tesco_is_doing/carbon_labelling.page.
23. H. Gujba and A. Azapagic, *The Carbon Footprint of Drinks Packaging in the UK*, The University of Manchester, UK, 2010.

Subject Index